Michael Laska

Elliptic Curves over Number Fields with Prescribed Reduction Type

Aspects of Mathematics
Aspekte der Mathematik

Editor: Klas Diederich

The texts published in this series are intended for graduate students and all mathematicians who wish to broaden their research horizons or who simply want to get a better idea of what is going on in a given field. They are introductions to areas close to modern research at a high level and prepare the reader for a better understanding of research papers. Many of the books can also be used to supplement graduate course programs.

The series will comprise two sub-series, one with English texts only and the other in German.

Michael Laska

Elliptic Curves over Number Fields with Prescribed Reduction Type

Friedr. Vieweg & Sohn   Braunschweig/Wiesbaden

CIP-Kurztitelaufnahme der Deutschen Bibliothek

Laska, Michael:
Elliptic curves over number fields with prescribed
reduction type/Michael Laska. — Braunschweig;
Wiesbaden: Vieweg, 1983.
    (Aspects of mathematics; Vol. 4)

NE: GT

Dr. *Michael Laska* is research fellow at the Max-Planck-Institut für Mathematik,
Bonn.

1983

Produced by IVD, Walluf b. Wiesbaden

ISBN 978-3-528-08569-8   ISBN 978-3-322-87599-0 (eBook)
DOI 10.1007/978-3-322-87599-0

- v -

Contents

Acknowledgements

I wish to express my gratitude to the Deutsche Forschungsgemeinschaft for financial support (Grant No. 468/1-2) during the preparation of this book.

I should also like to thank the Max-Planck-Institut für Mathematik at Bonn for its hospitality while this book was written.

Finally I am grateful to Professor Fritz Grunewald for many helpful comments.

Bonn, June 1983   Michael Laska

For Monika and Lukas

Introduction

Let K be an algebraic number field. The function attaching to each elliptic curve over K its conductor is constant on isogeny classes of elliptic curves over K (for the definitions see chapter 1). Moreover, for a given ideal $\mathfrak{a}$ in $\mathcal{O}_K$ the number of isogeny classes of elliptic curves over K with conductor $\mathfrak{a}$ is finite. In these notes we deal with the following <u>problem:</u> How can one explicitly construct a set of representatives for the isogeny classes of elliptic curves over K with conductor $\mathfrak{a}$ for a given ideal $\mathfrak{a}$ in $\mathcal{O}_K$?

The conductor of an elliptic curve over K is a numerical invariant which measures, in some sense, the badness of the reduction of the elliptic curve modulo the prime ideals in $\mathcal{O}_K$. It plays an important role in the famous Weil-Langlands conjecture on the connection between elliptic curves over K and congruence subgroups in $SL_2(\mathcal{O}_K)$. In case $K = \mathbb{Q}$ this connection can be stated as follows. For any ideal $\mathfrak{a} = (N)$ in $\mathbb{Z}$ let $\Gamma_0(N)$ be the congruence subgroup

$$\Gamma_0(N) = \left\{ \begin{pmatrix} a & b \\ c & d \end{pmatrix} \in SL_2(\mathbb{Z}) : c \in (N) \right\}$$

of $SL_2(\mathbb{Z})$ and let $S_2(\Gamma_0(N))$ be the space of cusp forms of weight 2 for $\Gamma_0(N)$. Now Weil conjectured that there exists a bijection between the rational normalized eigenforms in $S_2(\Gamma_0(N))$ for the Heckealgebra and the isogeny classes of elliptic curves over $\mathbb{Q}$ with conductor $\mathfrak{a} = (N)$. This bijection should have the property that the L-series $L(f,s)$ of a cusp form f coincides with the L-series $L(E,s)$ of the elliptic curve E over $\mathbb{Q}$, where E is a representative in the isogeny class corresponding to the form f . For more detailed informations on this connection we refer the reader to [Bi&Swi 2] or [Ge] . An analogue of Weil's conjecture over $\mathbb{Q}$ for imaginary quadratic number fields was investigated by Grunewald and Mennicke and can be found in [El&Gru&Me] or [Gru&Me] . It is now obvious that a solution to our above problem would be of considerable interest for an experimental inspection of the Weil-Langlands conjecture over $\mathbb{Q}$ resp. its analogue over other number fields.

For a given ideal $\mathfrak{a}$ in O_K , not only the number of isogeny classes of elliptic curves over K with conductor $\mathfrak{a}$ is finite, but also the number of curves itself. So, for a given ideal $\mathfrak{a}$ in O_K , one first of all would like to construct the elliptic curves over K with conductor $\mathfrak{a}$ and then find the division into the isogeny classes. However, much more is true. Indeed, if S is any given finite set of prime ideals in O_K then, by a theorem of Shafarevich (see [Lan 1]), there exists only a finite number of elliptic curves over K with conductor divisible in O_K only by prime ideals in S . These are the so-called elliptic curves over K with good reduction outside S . So, finally, for any given such S we will proceed in three steps.

(1) Explicit construction of all elliptic curves over K with good re-
     duction outside S .

(2) Determination of the conductor of each of these curves, i.e. deter-

mination of the exponent of the conductor at each prime in S .

(3)   Sorting the curves according to their conductor and, for each such

      conductor, finding the division into the isogeny classes.

These three steps will provide a solution to our above problem for any ideal

$\mathfrak{a}$ in O_K divisible in O_K only by primes in S . Our key tool in dealing

with the first step are diophantine methods and we will see that the con-

struction of all elliptic curves over K with good reduction outside S

may indeed be formulated as a certain diophantine problem over K . From

Baker's effectivity result on linear forms of logarithms of algebraic numbers

(see [Coa]) it follows that this diophantine problem is algorithmically

solvable. Baker's result, however, can in general not be used for explicit

calculations. Now in these notes we have the "practical" implementation

of the above steps in mind.

For a good understanding of these notes we assume that the reader is

familiar with the foundations of algebraic number theory. In chapter 1

we recall the notion of reduction of elliptic curves. In particular we

mention an algorithm by which the second of our above three steps can be

carried out. Chapters 2 and 3 deal with the first of our three steps, where

we assume that O_K is a principal ideal domain. In chapter 2 we show how

to to explicitly construct the elliptic curves over K with good reduction

outside some finite set S of primes in O_K from solutions of a certain

exponential diophantine equation over O_K , whereas in chapter 3 we deal

with the explicit construction of the solutions of the diophantine equation

themselves. This latter construction is equivalent to the study of a certain

finite number of equations of type $x^3 - y^2 = r$ with solutions in the

S-arithmetic subring $O_K[S^{-1}]$ of K . Chapter 4 deals with the last of our three steps. We show how the division into isogeny classes for the set of elliptic curves over K with given conductor can in many cases effectively be carried out. We further show how such classes can be visualized by a certain type of graphs. The diophantine problem we are dealing with in these notes was established in several special situations arising from concrete choices of K and S . In each case the solution of that problem leads to an explicit construction of the elliptic curves over K with good reduction outside S . Such results may be found in works of Coghlan, Ogg, Stroeker and the author. In chapter 5 we give a review on these and related results. The appendix contains the collection of tables of elliptic curves for the specific choice $K = \mathbb{Q}(i)$, $S = \{1+i,3\}$, which was obtained from the author's thesis and subsequent calculations ([Las 2], [Las 4]).

Throughout these notes we keep the algebraic number field K fixed. For K we use the following standard <u>notations</u>.

O_K is the ring of integers of K .

$\{\omega_1,\omega_2\}$ is an integral basis for K (K quadratic).

$\|\mathfrak{p}\|$ is the norm over Q of the prime ideal $\mathfrak{p}$ of O_K .

$\nu_{\mathfrak{p}}$ is the $\mathfrak{p}$-adic valuation of K corresponding to the prime ideal $\mathfrak{p}$ of O_K .

$R_{\mathfrak{p}}$ is the valuation ring of $\nu_{\mathfrak{p}}$ with maximal ideal $m_{\mathfrak{p}}$.

$k_{\mathfrak{p}}$ is the residue class field $R_{\mathfrak{p}}/m_{\mathfrak{p}}$.

O_K^* , $R_{\mathfrak{p}}^*$ is the group of units in O_K , $R_{\mathfrak{p}}$.

$W_K = \langle\zeta\rangle$ is the group of roots of unity in K with fixed generator ζ .

η is a fundamental unit in K (K real quadratic, otherwise put $\eta = 1$).

$\overline{K}$ is an algebraic closure of K .

G   is the absolute Galoisgroup $\mathrm{Gal}(\overline{K},K)$ over K .

A list of special symbols is given at the end of these notes.

Chapter 1

Reduction of elliptic curves

In this chapter we state the basic facts about reduction of elliptic curves over K . More detailed informations about elliptic curves may be obtained, for example, from Tate's [Ta 1] or Stroeker's [Stro 2] article.

An _elliptic curve over K_ is by definition a projective non-singular algebraic curve over K of genus 1 , furnished with a K-rational point P . Every such curve has a plane affine model of the form

$$\Gamma : \quad y^2 + a_1 xy + a_3 y = x^3 + a_2 x^2 + a_4 x + a_6$$

with coefficients a_i in K . In the projective plane $\mathbb{P}^2(\overline{K})$ the K-rational point P corresponds to the unique point $\underline{0} = (0,1,0)$ of Γ at infinity. Γ is called a (generalized) _Weierstrass equation_ for E over K . Let E and E' be elliptic curves over K , E with Weierstrass equation Γ as above, E' with Weierstrass equation

$$\Gamma' : \quad y'^2 + a_1' x'y' + a_3'y' = x'^3 + a_2'x'^2 + a_4'x' + a_6'$$

with coefficients $a_i' \in K$. By definition E and E' are _isomorphic_ (over K), if there exists an isomorphism λ of curves from E to E' such that

λ and λ^{-1} are defined over K and such that the distinguished K-rational points P and P' on E resp. E' are carried onto each other. In terms of the Weierstrass equations the isomorphy amounts to the following. E and E' are isomorphic (over K) if and only if there is a coordinate change of the form

$$x = u^2 x' + r$$
$$y = u^3 y' + u^2 s x' + t$$

with $r, s, t \in K$ and $u \in K*$. If this is the case, the coefficients of the equation Γ are related to the coefficients of the equation Γ' by the following formulas:

$$
\begin{aligned}
u a_1' &= a_1 + 2s \\
u^2 a_2' &= a_2 - s a_1 + 3r - s^2 \\
u^3 a_3' &= a_3 + r a_1 + 2t \\
u^4 a_4' &= a_4 - s a_3 + 2r a_2 - (t+rs)a_1 + 3r^2 - 2st \\
u^6 a_6' &= a_6 + r a_4 + r^2 a_2 + r^3 - t a_3 - t^2 - t r a_1 \quad .
\end{aligned}
$$

(1.1)

This is easily checked. We consider the set of all equations Γ of the form $\Gamma : y^2 + a_1 xy + a_3 y = x^3 + a_2 x^2 + a_4 x + a_6$ with coefficients a_i in K . Two such equations are called <u>equivalent</u> (over K) if they can be obtained from each other by means of a coordinate change of the above type. We define K-valued functions c_4, c_6, Δ, j on this set of equations satisfying certain homogenity properties. Following Tate [Ta 1] we first define functions b_2, b_4, b_6, b_8 in the following way:

$$
\begin{aligned}
b_2(\Gamma) &= a_1^2 + 4a_2 \\
b_4(\Gamma) &= a_1 a_3 + 2a_4
\end{aligned}
$$

(1.2)

$$(1.2) \qquad b_6(\Gamma) = a_3^2 + 4a_6$$

$$b_8(\Gamma) = a_1^2 a_6 - a_1 a_3 a_4 + 4a_2 a_6 + a_2 a_3^2 - a_4^2 \quad .$$

Now the functions c_4, c_6, Δ, j are defined as follows:

$$(1.3) \qquad
\begin{aligned}
c_4 &= b_2^2 - 24b_4 \\
c_6 &= -b_2^3 + 36b_2 b_4 - 216b_6 \\
\Delta &= -b_2^2 b_8 - 8b_4^3 - 27b_6^2 + 9b_2 b_4 b_6 \\
j &= c_4^3 / \Delta \quad .
\end{aligned}$$

Note that $\Delta(\Gamma) \neq 0$ if and only if Γ is a model of an elliptic curve over K or, what amounts to the same thing, the plane curve defined by the equation Γ is non-singular. The functions c_4, c_6, Δ are algebraic dependent. Indeed, we have

$$c_4^3 - c_6^2 = 2^6 3^3 \Delta \quad .$$

Let Γ, Γ' be equivalent equations so that there exists a coordinate change $x = u^2 x' + r$, $y = u^3 y' + u^2 s x' + t$ with $r,s,t \in K$ and $u \in K^*$ taking Γ to Γ' . Let $b_i = b_i(\Gamma)$, $c_\ell = c_\ell(\Gamma)$, $\Delta = \Delta(\Gamma)$, $j = j(\Gamma)$ and similarly $b_i' = b_i(\Gamma')$ etc., $i = 2,4,6,8$ and $\ell = 4,6$. Then we have

$$(1.4) \qquad
\begin{aligned}
u^2 b_2' &= b_2 + 12r \\
u^4 b_4' &= b_4 + r b_2 + 6r^2 \\
u^6 b_6' &= b_6 + 2r b_4 + r^2 b_2 + 4r^3 \\
u^8 b_8' &= b_8 + 3r b_6 + 3r^2 b_4 + r^3 b_2 + 3r^4 \\[1em]
u^4 c_4' &= c_4 \qquad\qquad u^{12} \Delta' = \Delta \\
u^6 c_6' &= c_6 \qquad\qquad\quad j' = j \quad .
\end{aligned}$$

We see that the condition $\Delta(\Gamma) \neq 0$ is invariant for the equivalence class containing Γ and moreover, j is invariant on each such class. We write $j(E) = j(\Gamma)$, if Γ is any Weierstrass equation over K for the elliptic curve E over K. $\Delta(\Gamma)$ is called the <u>discriminant</u> of Γ and $j(E)$ the <u>j-invariant</u> of E.

The equation $\Gamma : y^2 + a_1 xy + a_3 y = x^3 + a_2 x^2 + a_4 x + a_6$ with all a_i in K is equivalent to the equation

$$(1.5) \qquad \Gamma^{(u)} : \quad y'^2 = x'^3 - \frac{c_4(\Gamma)}{48u^4} x' - \frac{c_6(\Gamma)}{864u^6} \quad ,$$

where $u \in K*$ is arbitrary. Indeed, for $u \in K*$ we may obtain $\Gamma^{(u)}$ from Γ by means of the coordinate change $x = u^2 x' + r$, $y = u^3 y' + u^2 s x' + t$ with $r = -\frac{1}{12} b_2(\Gamma)$, $s = -\frac{1}{2} a_1$, $t = -\frac{1}{2} a_3 + \frac{1}{24} a_1 b_2(\Gamma)$. We have $\Delta(\Gamma) = u^{12} \Delta(\Gamma^{(u)})$. An important choice for u will be $u = 1/6$.

(1.6) <u>Definition.</u> Let E be an elliptic curve over K . Let $\mathfrak{p}$ be a prime ideal in O_K . A Weierstrass equation $\Gamma : y^2 + a_1 xy + a_3 y = x^3 + a_2 x^2 + a_4 x + a_6$ for E over K is called <u>minimal at $\mathfrak{p}$</u> if all $a_i \in R_{\mathfrak{p}}$ and $v_{\mathfrak{p}}(\Delta(\Gamma))$ is minimal subject to that condition.

(1.7) <u>Lemma.</u> Let E be an elliptic curve over K . Let $\mathfrak{p}$ be a prime ideal in O_K . Then there exists a Weierstrass equation Γ for E over K such that Γ is minimal at $\mathfrak{p}$. Any other Weierstrass equation for E over K , minimal at $\mathfrak{p}$, may be obtained from Γ by means of a coordinate change of the form $x = u^2 x' + r$, $y = u^3 y' + u^2 s x' + t$ with $r, s, t \in R_{\mathfrak{p}}$, $u \in R_{\mathfrak{p}}^*$.

<u>Proof.</u> For the existence let Γ' : $y'^2 + a_1' x' y' + a_3' y' = x'^3 + a_2' x'^2 + a_4' x' + a_6'$ be any Weierstrass equation for E over K . Choose $u \in K^*$ such that $v_{\mathfrak{p}}(u) \geq -\frac{1}{i} v_{\mathfrak{p}}(a_i')$ for all i . This yields an equation Γ for E over K with coefficients a_i , where $a_i = u^i a_i'$. Hence $v_{\mathfrak{p}}(a_i) \geq 0$ and $a_i \in R_{\mathfrak{p}}$ for all i . It follows $v_{\mathfrak{p}}(\Delta(\Gamma)) \geq 0$. Since $v_{\mathfrak{p}}$ is a discrete valuation on K there exists an equation Γ with coefficients in $R_{\mathfrak{p}}$ such that $v_{\mathfrak{p}}(\Delta(\Gamma))$ is minimal subject to all equations for E over K with coefficients in $R_{\mathfrak{p}}$. This proves the existence. For the uniqueness let Γ and Γ' be two Weierstrass equations for E over K both minimal at $\mathfrak{p}$ with coefficients a_i and a_i' resp. in $R_{\mathfrak{p}}$. Let $x = u^2 x' + r$, $y = u^3 y' + u^2 s x' + t$ with $r, s, t \in K$ and $u \in K^*$ be a coordinate change taking Γ to Γ' . Since $v_{\mathfrak{p}}(\Delta(\Gamma)) = v_{\mathfrak{p}}(\Delta(\Gamma'))$ we conclude from $\Delta(\Gamma) = u^{12}\Delta(\Gamma')$ (see 1.4) that $u \in R_{\mathfrak{p}}^*$. We have $b_i(\Gamma)$, $b_i(\Gamma') \in R_{\mathfrak{p}}$ for all i . From the equations for $u^6 b_6'$ and $u^8 b_8'$ in 1.4 we conclude that $4r$ and $3r$ resp. are zeroes of monic polynomials over $R_{\mathfrak{p}}$. Since $R_{\mathfrak{p}}$ is a principal ideal domain and $4r, 3r \in K$ we see that $4r, 3r \in R_{\mathfrak{p}}$ and hence $r \in R_{\mathfrak{p}}$. Next consider the equations for $u^2 a_2'$ and $u^6 a_6'$ in 1.1. From these equations we conclude that $s \in R_{\mathfrak{p}}$ and $t \in R_{\mathfrak{p}}$ resp.. The Lemma is proved. ∎

Let E be an elliptic curve over K and let Γ be a Weierstrass equation for E over K with coefficients a_i . Put $c_4 = c_4(\Gamma)$, $c_6 = c_6(\Gamma)$, $\Delta = \Delta(\Gamma)$. Let $\mathfrak{p}$ be a prime ideal in O_K . If $a_i \in R_{\mathfrak{p}}$ and $v_{\mathfrak{p}}(\Delta) < 12$, then Γ is clearly minimal at $\mathfrak{p}$ by 1.4. Conversely, suppose that Γ is minimal at $\mathfrak{p}$. Suppose that $j(E) \in R_{\mathfrak{p}}$. Then

$$v_{\mathfrak{p}}(\Delta) < 12 + 12 v_{\mathfrak{p}}(2) + 6 v_{\mathfrak{p}}(3) \quad .$$

Indeed, let $\pi \in R_{\mathfrak{p}}$ be a prime such that $v_{\mathfrak{p}}(\pi) = 1$. Consider the equation $\Gamma' = \Gamma^{(\pi)}$ for E with coefficients $a_4' = \dfrac{c_4(\Gamma)}{48\pi^4}$, $a_6' = \dfrac{c_6(\Gamma)}{864\pi^6}$ (see 1.5). We have $\Delta = \pi^{12}\Delta(\Gamma')$. Since Γ is minimal at π , we must have $a_4' \notin R_{\mathfrak{p}}$ or $a_6' \notin R_{\mathfrak{p}}$. The condition $j(E) \in R_{\mathfrak{p}}$ means that $v_{\mathfrak{p}}(\Delta) \leq 3v_{\mathfrak{p}}(c_4)$. Now suppose that $a_4' \notin R_{\mathfrak{p}}$. Then $v_{\mathfrak{p}}(\Delta) \leq 3v_{\mathfrak{p}}(4) < 12+12v_{\mathfrak{p}}(2)+3v_{\mathfrak{p}}(3)$. Suppose that $a_6' \notin R_{\mathfrak{p}}$. Then $\min\{v_{\mathfrak{p}}(c_4^3), v_{\mathfrak{p}}(2^6 3^3 \Delta)\} \leq 2v_{\mathfrak{p}}(c_6)$ since $c_4^3 - c_6^2 = 2^6 3^3 \Delta$. If $v_{\mathfrak{p}}(c_4^3)$ is the minimum, then $v_{\mathfrak{p}}(\Delta) \leq 3v_{\mathfrak{p}}(c_4) < 12+10v_{\mathfrak{p}}(2)+6v_{\mathfrak{p}}(3)$. If $v_{\mathfrak{p}}(2^6 3^3 \Delta)$ is the minimum, then $v_{\mathfrak{p}}(\Delta) \leq 12+4v_{\mathfrak{p}}(2)+3v_{\mathfrak{p}}(3)$. In any case one obtains $v_{\mathfrak{p}}(\Delta) < 12+12v_{\mathfrak{p}}(2)+6v_{\mathfrak{p}}(3)$ as claimed.

If Γ is a Weierstrass equation over K for an elliptic curve over K , say with coefficients a_i , then the condition $v_{\mathfrak{p}}(a_i) \geq 0$ for all i fails only for a finite number of prime ideals $\mathfrak{p}$ in $\mathcal{O}_K$. Thus generalizing the argument for the proof of existence in Lemma 1.7 we see that each elliptic curve over K has a Weierstrass equation with coefficients in $\mathcal{O}_K = \underset{\mathfrak{p}}{\cap} R_{\mathfrak{p}}$. Such a Weierstrass equation is called an <u>integral</u> Weierstrass equation (over K).

For an elliptic curve E over K and a prime ideal $\mathfrak{p}$ in $\mathcal{O}_K$ we put

$$d_{\mathfrak{p}} = v_{\mathfrak{p}}(\Delta(\Gamma)) \quad ,$$

where Γ is some Weierstrass equation for E over K , minimal at $\mathfrak{p}$.

(1.8) <u>Definition.</u> Let E be an elliptic curve over K , $\mathfrak{p}$ a prime ideal in $\mathcal{O}_K$. The ideal $\mathfrak{p}^{d_{\mathfrak{p}}}$ is called the <u>discriminant at $\mathfrak{p}$</u> of E .

The ideal

$$\mathrm{Disc}(E) = \prod_{\mathfrak{p}} \mathfrak{p}^{d_{\mathfrak{p}}}$$

is called the <u>discriminant</u> of E .

Clearly, the discriminant of E divides any principal ideal $(\Delta(\Gamma))$, where Γ is an integral Weierstrass equation for E over K . Moreover, if $\mathfrak{p}$ is any prime ideal in O_K , then $\nu_{\mathfrak{p}}(\Delta(\Gamma)) = 12\nu_{\mathfrak{p}}(u_{\mathfrak{p}})+d_{\mathfrak{p}}$ for some suitable $u_{\mathfrak{p}} \in R_{\mathfrak{p}}$ (see the definition of $d_{\mathfrak{p}}$ and the formulas 1.4). Thus we conclude that

$$(\Delta(\Gamma)) = \mathfrak{a}^{12}\,\mathrm{Disc}(E) \qquad ,$$

for some ideal $\mathfrak{a}$ in O_K .

<u>(1.9) Definition.</u> Let E be an elliptic curve over K . An integral Weierstrass equation Γ for E over K is called <u>global minimal</u> if $\mathrm{Disc}(E) = (\Delta(\Gamma))$.

A necessary condition for the existence of a global minimal Weierstrass equation over K for an elliptic curve E over K is that $\mathrm{Disc}(E)$ is a principal ideal in O_K . We show that in case $(h_K,6) = 1$ the converse is always true.

(1.10) **Proposition.** Suppose the class number of K is prime to 6. Let E be an elliptic curve over K such that the discriminant of E is a principal ideal in O_K. Then E has a global minimal Weierstrass equation over K.

Proof. Let Γ be any integral Weierstrass equation for E over K with coefficients, say a_i. We have $(\Delta(\Gamma)) \operatorname{Disc}(E)^{-1} = \mathfrak{a}^{12}$ for some ideal $\mathfrak{a}$ in O_K. In particular, by our assumption on $\operatorname{Disc}(E)$, $\mathfrak{a}^{12}$ is a principal ideal. The condition on the class number of K then implies that $\mathfrak{a}$ itself is a principal ideal in O_K, say $\mathfrak{a} = (u)$, $u \in O_K$. Let $\mathfrak{p}$ be any prime ideal in O_K and let $\Gamma_\mathfrak{p}$ be a Weierstrass equation for E over K, minimal at $\mathfrak{p}$, say with coefficients $a_i^{(\mathfrak{p})} \in R_\mathfrak{p}$. With u as above and with a_i' as $a_i^{(\mathfrak{p})}$ we solve the equations in 1.1 successively for $s = s_\mathfrak{p}$, $r = r_\mathfrak{p}$, $t = t_\mathfrak{p}$. This gives a transformation $x = u^2 x' + r_\mathfrak{p}$, $y = u^3 y' + u^2 s_\mathfrak{p} x' + t_\mathfrak{p}$ carrying Γ to $\Gamma_\mathfrak{p}$. Now for each $\mathfrak{p}$ the 5 simultaneous congruences $a_1 + 2s \equiv 0 \bmod u$, $a_2 - sa_1 + 3r - s^2 \equiv 0 \bmod u^2$, etc. have a solution $(r_\mathfrak{p}, s_\mathfrak{p}, t_\mathfrak{p})$ in $R_\mathfrak{p}^3$. By the Chinese remainder Theorem one can construct a global solution (r,s,t) in K^3. The transformation $x = u^2 x' + r$, $y = u^3 y' + u^2 s x' + t$ carries Γ to an integral equation Γ' for E over K. Moreover, since $(u^{12})(\Delta(\Gamma')) = (\Delta(\Gamma)) = (u^{12}) \operatorname{Disc}(E)$ we have $(\Delta(\Gamma')) = \operatorname{Disc}(E)$ and hence Γ' is a global minimal Weierstrass equation for E over K, as desired. ∎

The assumptions in Proposition 1.10 are obviously satisfied, for example, in the following situations: (i) K has class number 1 or (ii) the class number of K is prime to 6 and $\operatorname{Disc}(E)$ is the unit ideal. An algorithm which leads from an arbitrary integral Weierstrass equation of

an elliptic curve E over K to a global minimal Weierstrass equation for
E over K , provided such a global minimal equation does exist, is described
in [Las 1]. The assumption on the class number in Proposition 1.10 is really
necessary as the following example shows.

(1.11) <u>Example.</u> Consider the field $K = \mathbb{Q}(\sqrt{-5})$ with class number 2 .
Let E be the elliptic curve over K defined by the Weierstrass equation

$$\Gamma : \quad y^2 = x^3 - 4x \qquad .$$

We leave it as an exercise to the reader to show that E has no global
minimal Weierstrass equation over K . (Hint: (1) Show that the unique prime
divisor $\mathfrak{p} = (2, 1+\sqrt{-5})$ of 2 is the only prime ideal in O_K at which Γ
is not minimal. (2) Show that $y^2 = x^3 - Ax$ with $A = 4(1+\sqrt{-5})^{-4}$ is a
Weierstrass equation for E over K minimal at $\mathfrak{p}$.)

Let E be an elliptic curve over K . For each prime ideal $\mathfrak{p}$ in
O_K we are going to attach to E an elliptic curve $E_{\mathfrak{p}}$, called the
<u>reduction of E at $\mathfrak{p}$</u> . To do this we fix a Weierstrass equation Γ for
E over K such that Γ is minimal at $\mathfrak{p}$. The reduction of the coefficients of Γ modulo $\mathfrak{p}$ gives an equation $\overline{\Gamma}$ for a plane cubic curve defined over the finite field $k_{\mathfrak{p}}$. We denote this curve by

$$E_{\mathfrak{p}}$$

From Lemma 1.7 we conclude that $E_{\mathfrak{p}}$ is uniquely determined by $\overline{\Gamma}$ up to a
coordinate change of the form $x = u^2 x' + r$, $y = u^3 y' + u^2 s x' + t$ with $r, s, t \in$

$k_{\mathfrak{p}}$ and $u \in k_{\mathfrak{p}}^*$. Let $E_{\mathfrak{p}}(k_{\mathfrak{p}})$ be the vanishing set of $\overline{\Gamma}$ in the projective space $\mathbb{P}^2(k_{\mathfrak{p}})$. The set $E_{\mathfrak{p}}(k_{\mathfrak{p}})^{ns}$ of non-singular points of $E_{\mathfrak{p}}$ in $E_{\mathfrak{p}}(k_{\mathfrak{p}})$ can be given the structure of a commutative group in the following way: First of all, if $P_1, P_2 \in E_{\mathfrak{p}}(k_{\mathfrak{p}})^{ns}$ then the line through P_1 and P_2 (which is tangent in case $P_1 = P_2$) intersects the curve $E_{\mathfrak{p}}$ in a uniquely determined third point P_3 which is easily seen to be in $E_{\mathfrak{p}}(k_{\mathfrak{p}})^{ns}$. Now define $-P_3 := P_1 + P_2$. We have the following possibilities for $E_{\mathfrak{p}}$:

(1) $E_{\mathfrak{p}}$ is an <u>elliptic curve</u> over $k_{\mathfrak{p}}$. In this case $\Delta(\Gamma) \in R_{\mathfrak{p}}^*$ and $E_{\mathfrak{p}}(k_{\mathfrak{p}}) = E_{\mathfrak{p}}(k_{\mathfrak{p}})^{ns}$. We have $j(E) \in R_{\mathfrak{p}}$ and $j(E_{\mathfrak{p}}) = \overline{j(E)} \in k_{\mathfrak{p}}$. Moreover,

$$|E_{\mathfrak{p}}(k_{\mathfrak{p}})| = 1 + \|\mathfrak{p}\| - a \quad \text{with} \quad |a| \leq 2\sqrt{\|\mathfrak{p}\|} \quad .$$

(2) $E_{\mathfrak{p}}$ is a <u>rational curve</u> over $k_{\mathfrak{p}}$. In this case $\Delta(\Gamma) \in m_{\mathfrak{p}}$ and $E_{\mathfrak{p}}(k_{\mathfrak{p}})$ contains exactly one singular point $S_{\mathfrak{p}}$. $S_{\mathfrak{p}}$ is either a cusp or a node

(2a) If $S_{\mathfrak{p}}$ is a <u>cusp</u>, then $c_4(\Gamma) \in m_{\mathfrak{p}}$ and we have

$$E_{\mathfrak{p}}(k_{\mathfrak{p}})^{ns} \simeq k_{\mathfrak{p}}^+ \quad , \quad |E_{\mathfrak{p}}(k_{\mathfrak{p}})| = \|\mathfrak{p}\| + 1 \quad .$$

(2b) If $S_{\mathfrak{p}}$ is a <u>node</u>, then $c_4(\Gamma) \in R_{\mathfrak{p}}^*$ and $j(E) \notin R_{\mathfrak{p}}$ and we have

$$E_{\mathfrak{p}}(k_{\mathfrak{p}})^{ns} \simeq k_{\mathfrak{p}}^* \quad , \quad |E_{\mathfrak{p}}(k_{\mathfrak{p}})| = \|\mathfrak{p}\| \quad ,$$

provided the two tangents to $E_{\mathfrak{p}}$ at $S_{\mathfrak{p}}$ are defined over $k_{\mathfrak{p}}$. If they are not, then they are defined over the quadratic extension L of $k_{\mathfrak{p}}$ and in this case we have

$$E_{\mathfrak{p}}(k_{\mathfrak{p}})^{ns} \simeq \{x \in L \mid N_{L:k_{\mathfrak{p}}}(x) = 1\} \subseteq L^* \quad , \quad |E_{\mathfrak{p}}(k_{\mathfrak{p}})| = \|\mathfrak{p}\| + 2 \quad .$$

(1.12) Definition. Let E be an elliptic curve over K and let $\mathfrak{p}$ be a prime ideal in O_K . E is said to have <u>good reduction</u> at $\mathfrak{p}$, if $E_{\mathfrak{p}}$ is an elliptic curve, <u>bad reduction</u> at $\mathfrak{p}$ otherwise. In particular, if $S_{\mathfrak{p}}$ is a node, E is said to have <u>multiplicative</u>, or <u>semistable</u>, reduction at $\mathfrak{p}$ and, if $S_{\mathfrak{p}}$ is a cusp, E is said to have <u>additive</u>, or <u>unstable</u>, re- duction at $\mathfrak{p}$. E is said to have <u>potential good reduction</u> at $\mathfrak{p}$, if for some finite extension field K' of K the curve E over K' has good reduction at $\mathfrak{p}'$ for any prime divisor $\mathfrak{p}'$ of $\mathfrak{p}$ in $O_{K'}$.

It can be shown (see [Deu]) that E has potential good reduction at $\mathfrak{p}$ if and only if $j(E) \in R_{\mathfrak{p}}$. By checking all the above possibilities for $E_{\mathfrak{p}}$ we see that the condition $j(E) \in R_{\mathfrak{p}}$ is equivalent to the condition that E has either good reduction at $\mathfrak{p}$ or additive reduction at $\mathfrak{p}$ with $j(E) \in R_{\mathfrak{p}}$. If E has additive reduction at $\mathfrak{p}$ with $j(E) \notin R_{\mathfrak{p}}$, then there exists a finite extension field K' of K such that E over K' has multiplicative reduction at each prime divisor $\mathfrak{p}'$ of $\mathfrak{p}$ in $O_{K'}$. Clearly, E has bad reduction at precisely those finitely many prime ideals in O_K dividing the discriminant of E .

We can now define the L-series of an elliptic curve E over K . For this we put

$$a_{\mathfrak{p}} = \|\mathfrak{p}\| + 1 - |E_{\mathfrak{p}}(k_{\mathfrak{p}})| \quad ,$$

for any prime ideal $\mathfrak{p}$ in $\mathcal{O}_K$. Form the above we conclude that

$$|a_{\mathfrak{p}}| \leq 2\sqrt{\|\mathfrak{p}\|} \quad ,$$

in case E has good reduction at $\mathfrak{p}$, and

$$a_{\mathfrak{p}} = \begin{cases} 0 & \text{, if } E \text{ has additive reduction at } \mathfrak{p} \text{ .} \\ 1 & \text{, if } E \text{ has multiplicative reduction at } \mathfrak{p} \\ & \quad \text{with tangents defined over } k_{\mathfrak{p}} \text{ .} \\ -1 & \text{, if } E \text{ has multiplicative reduction at } \mathfrak{p} \\ & \quad \text{with tangents not defined over } k_{\mathfrak{p}} \text{ .} \end{cases}$$

The <u>L-series</u> of E is now given by the Euler product

$$(1.13) \qquad L(E,s) = \prod_{\mathfrak{p}} \frac{1}{1 - a_{\mathfrak{p}} \|\mathfrak{p}\|^{-s} + \psi(\mathfrak{p}) \|\mathfrak{p}\|^{1-2s}} \quad ,$$

where $\psi(\mathfrak{p}) = 1$ if E has good reduction at $\mathfrak{p}$ and $\psi(\mathfrak{p}) = 0$ otherwise. The product converges for $\mathrm{Re}(s) > 3/2$. In particular the product determines a holomorphic function of s in the right half plane $\mathrm{Re}(s) > 3/2$. The <u>conjecture of Weil and Langlands</u> states that $L(E,s)$ coincides with the L-series of an appropriate automorphic form. A consequence of this conjecture is that $L(E,s)$ has an analytic continuation to the whole plane and satisfies a simple functional equation.

We also recall the reduction of an elliptic curve in the sense of Néron and Kodaira. Let E be an elliptic curve over K and let $\mathfrak{p}$ be a prime ideal in $\mathcal{O}_K$. There exists a unique (up to isomorphy) regular sheme X over $R_{\mathfrak{p}}$ such that $E = X \underset{R_{\mathfrak{p}}}{\times} K$ and such that $X \to \mathrm{Spec}\, R_{\mathfrak{p}}$

cannot be factored as $X \to X' \to \operatorname{Spec} R_{\mathfrak{p}}$ in such a way that $E = X' \underset{R_{\mathfrak{p}}}{\times} K$.
X is called the _Néron model_ of E at $\mathfrak{p}$. Its fibre $X_{\mathfrak{p}} = X \underset{R_{\mathfrak{p}}}{\times} k_{\mathfrak{p}}$ is called
the _Néron-Kodaira reduction_ of E at $\mathfrak{p}$. $X_{\mathfrak{p}}$ is one of the 10 types

$$I_0 \ , \ I_\nu(\nu>0) \ , \ II \ , \ III \ , \ IV \ , \ I_0^* \ , \ I_\nu^*(\nu>0) \ , \ IV^* \ , \ III^* \ , \ II^* \quad ,$$

listed in [Ta 2] . Let $n_{\mathfrak{p}}$ be the total number of irreducible components
(not counting multiplicities) of $X_{\mathfrak{p}}$. Let $d_{\mathfrak{p}}$ be the exponent of the
discriminant of E at $\mathfrak{p}$. Put

$$f_{\mathfrak{p}} = d_{\mathfrak{p}} + 1 - n_{\mathfrak{p}} \quad .$$

Then Ogg has shown that (see [Ogg 1])

$$f_{\mathfrak{p}} = \begin{cases} 0 & \text{, if } E \text{ has good reduction at } \mathfrak{p} \text{ .} \\ 1 & \text{, if } E \text{ has multiplicative reduction at } \mathfrak{p} \text{ .} \\ 2+\delta \ , \ \delta \geq 0 & \text{, if } E \text{ has additive reduction at } \mathfrak{p} \text{ , with} \\ & \quad \delta = 0 \text{ if } \mathfrak{p} \nmid 2,3 \text{ .} \end{cases}$$

(1.14) **Definition.** Let E be an elliptic curve over K . For a prime
ideal $\mathfrak{p}$ in 0_K the ideal $\mathfrak{p}^{f_{\mathfrak{p}}}$ is called the _conductor of E at $\mathfrak{p}$_ .
The numerical invariant

$$\operatorname{Cond}(E) = \prod_{\mathfrak{p}} \mathfrak{p}^{f_{\mathfrak{p}}}$$

is called the _conductor_ of E (over K).

Note that Cond(E) divides Disc(E) and that both Cond(E) and Disc(E) are exactly divisible by those primes in O_K at which E has bad reduction. The type of $X_{\mathfrak{p}}$ and in particular $f_{\mathfrak{p}}$ can be determined from an arbitrary Weierstrass equation for E over K by means of the Tate algorithm, described in [Ta 2] . This corresponds to step (2) in the introduction. The importance of the conductor of an elliptic curve was pointed out in the introduction. In the following two chapters we deal with the explicit construction of all those elliptic curves over K , whose conductor have a certain shape.

Chapter 2

Elliptic curves with good reduction outside
a given set of prime ideals

In this chapter we give an explicit parametrization in terms of a certain diophantine equation over K of all elliptic curves over K with good reduction outside a given finite set of prime ideals in O_K .

If S is a set of prime ideals in O_K we let

$$E(S)$$

be the set of all elliptic curves over K with good reduction at all primes of O_K not in S (we simply say: with good reduction outside S). The case $S = \emptyset$ is included. In this case $E(S)$ is the set of elliptic curves over K with good reduction everywhere. We can now formulate the **main subject of this and the following chapter:** The explicit construction of all elliptic curves in $E(S)$ for any given finite set S of prime ideals in O_K as the solution of a certain diophantine problem over K . This corresponds to step (1) in the introduction. By definition of the conductor (see 1.14), the curves in $E(S)$ are exactly those with conductor divisible in O_K only by primes in S . The reduction type and the actual exponent of the conductor at each prime in S may be obtained by means of the

Tate algorithm, described in [Ta 2].

In the following we fix a finite set S of prime ideals in O_K , where for technical reasons we assume that <u>all prime ideals dividing 2 or 3 are contained in S</u> . The case of good reduction at these primes will be discussed separately. A simple criterion for an elliptic curve over K to be in $E(S)$ is given in the following Lemma.

(2.1) <u>Lemma.</u> Let E be an elliptic curve over K . Suppose that E has a global minimal Weierstrass equation over K . Then E is in $E(S)$ if and only if E has a Weierstrass equation of the form

$$y^2 = x^3 - 27vx - 54w$$

with $v,w \in O_K$ such that if a prime $\mathfrak{p}$ in O_K divides $(v^3 - w^2)$, then $\mathfrak{p} \in S$.

<u>Proof.</u> Suppose that E has an equation $\Gamma : y^2 = x^3 - 27vx - 54w$ such that $v,w \in O_K$ with $\mathfrak{p} \nmid (v^3 - w^2)$ for all $\mathfrak{p} \notin S$. Then $\Delta(\Gamma) = 2^6 3^9 (v^3 - w^2)$. Hence $v_{\mathfrak{p}}(\Delta(\Gamma)) = 0$ for all $\mathfrak{p} \notin S$ and hence Γ is minimal at such $\mathfrak{p}$ and in particular E has good reduction outside S . Conversely, suppose that E has good reduction outside S and let

$$\Gamma : y^2 + a_1xy + a_3y = x^3 + a_2x^2 + a_4x + a_6$$

be a global minimal Weierstrass equation for E over K . Then $a_i \in O_K$

for all i , $\Delta(\Gamma) \in O_K$ and $v_{\mathfrak{p}}(\Delta(\Gamma)) = 0$ for all $\mathfrak{p} \notin S$. We consider $\Gamma^{(u)}$ for $u = 1/6$ (see 1.5). Let $c_4 = c_4(\Gamma)$, $c_6 = c_6(\Gamma)$. Then

$$\Gamma^{(1/6)} : \quad y^2 = x^3 - 27c_4 x - 54c_6$$

is also an equation for E with $c_4, c_6 \in O_K$ such that $\mathfrak{p} \nmid 2^6 3^3 \Delta(\Gamma) = c_4^3 - c_6^2$ for all $\mathfrak{p} \notin S$. Hence E has an equation of the desired type. The Lemma is proved. ∎

For the rest of this chapter we assume that O_K is a <u>principal ideal domain</u>. In this case each elliptic curve over K has a global minimal Weierstrass equation over K by Proposition 1.10. From Lemma 2.1 we conclude that the construction of the elliptic curves in $E(S)$ is now connected with the solution of a certain diophantine problem over K . Indeed, let

$$S = \{\pi_1, \ldots, \pi_n\} \ ,$$

where the π_v are pairwise non-associated prime elements in O_K . Consider the exponential diophantine equation

$$(*) \qquad \boxed{\ x^3 - y^2 = \xi \pi_1^{z_1} \cdots \pi_n^{z_n}\ }$$

with solutions $(x, y, \xi, z_1, \ldots, z_n)$ such that $x, y \in O_K$, $\xi \in O_K^*$, $z_1, \ldots, z_n \in \mathbb{N}_0$. Then by Lemma 2.1 to each solution $(v, w, \varepsilon, e_1, \ldots, e_n)$ of equation $(*)$ one can associate an elliptic curve E in $E(S)$ with Weierstrass equation $y^2 = x^3 - 27vx - 54w$, and each elliptic curve in $E(S)$ arises in this way. In the following we study this connection more closely. In particular we give answers to the following questions:

(1) Which of the solutions of equation (*) define non-isomorphic curves?

(2) How can one obtain a global minimal Wierstrass equation for each
     such curve?

For the sake of simplicity we restrict our attention to the case, where
K is either <u>quadratic or the field $\mathbb{Q}$ of rational numbers</u>. Everything of
course is valid for higher degrees of K   (0_K principal ideal domain) , then,
however, the technical details become more complicated. For an answer to the
first question we introduce the following terminology.

$\underline{(2.2)}$ $\underline{\text{Definition.}}$ A solution $(p,q,\epsilon,m_1,\ldots,m_n)$ of equation (*) is
called a <u>basic solution</u> if the following condition is satisfied: If t^2
divides p and t^3 divides q for $t \in 0_K$ then $t \in 0_K^*$. Two such basic
solutions $(p,q,\epsilon,m_1,\ldots,m_n)$ and $(p',q',\epsilon',m_1',\ldots,m_n')$ are called
<u>associated</u>, if $p = \delta^2 p'$, $q = \delta^3 q'$ for some $\delta \in 0_K^*$.

In the following we fix a set $B = B_S$ of representatives of pairwise
non-associated basic solutions of equation (*). We will see in chapter 3
that B is in fact a finite set. All solutions of equation (*) are obvious-
ly of the form $(\lambda^2 p, \lambda^3 q, \epsilon\tau^6, m_1+6a_1,\ldots,m_n+6a_n)$, where $(p,q,\epsilon,m_1,\ldots,m_n)$
$\in B$ and $\lambda = \tau\pi_1^{a_1}\cdots\pi_n^{a_n}$ with $\tau \in 0_K^*$, $a_1,\ldots,a_n \in \mathbb{N}_0$. It is often
possible to choose a set B of representatives in a canonical way.

$\underline{(2.3)}$ $\underline{\text{Examples.}}$ (1) Let $K = \mathbb{Q}$. Define B as the set of all basic
solutions $(p,q,\epsilon,m_1,\ldots,m_n)$ of equation (*) such that $q \geq 0$. B is

clearly a set of representatives of non-associated basic solutions.

(2) Let $K = \mathbb{Q}(i)$. Define B as the set of all basic solutions $(p,q,\varepsilon,m_1,\ldots,m_n)$ of equation (*) such that $\mathrm{Re}(q) \geq 0$, where $\mathrm{Im}(q) \geq 0$ if $\mathrm{Re}(q) = 0$, and such that $\varepsilon \in \{1,i\}$. B is clearly a set of representatives of non-associated basic solutions.

For a basic solution $\beta = (p,q,\varepsilon,m_1,\ldots,m_n)$ we often simply write $\beta = (p,q)$. Note that $\varepsilon,m_1,\ldots,m_n$ are uniquely determined by (p,q) . Let $U = U_S = 0_K^* <S>$, where $<S>$ is the subgroup of K^* generated by the finite set S . Let $\beta = (p,q) \in B$ and let $\overline{\lambda} \in U/U^2$, $\lambda \in U$, where we can assume $\lambda \in 0_K$. Let E be the elliptic curve over K with Weierstrass equation

$$\Gamma : \quad y^2 = x^3 - 3\lambda^2 px - 2\lambda^3 q \quad .$$

Since $y^2 = x^3 - 3\mu^4\lambda^2 px - 2\mu^6\lambda^3 q$ with $\mu \in K^*$ is also an equation for E (see 1.1) we see that E is independent of the choice of the representative in the class of λ modulo U^2 . We denote E by $\underline{\beta*\overline{\lambda}}$ and define

$$\Phi(\beta,\lambda) = \beta*\overline{\lambda} \quad .$$

The following Theorem gives an answer to the first of our above two questions.

(2.4) __Theorem.__ $\Phi : B \times (U/U^2) \rightarrow E(S)$ is a bijection.

__Proof.__ We have $\Delta(\Gamma) = 2^6 3^3 \lambda^6 (p^3 - q^2)$ so that in fact $\beta*\overline{\lambda} \in E(S)$ by Lemma 2.1 . If $E \in E(S)$ is arbitrary, then again by Lemma 2.1 E has an equation of the form $y^2 = x^3 - 27vx - 54w$, where v,w can be written

as $v = \mu^2 p$, $w = \mu^3 q$ with some basic solution $\beta = (p,q)$ in B and

$\mu \in U$. Now S contains the primes in O_K dividing 3 and therefore, as

one can easily see, E has an equation of type $y^2 = x^3 - 3\lambda^2 px - 2\lambda^3 q$ for

a certain $\lambda \in U$. Thus we have $E = \beta * \overline{\lambda}$ and Φ is surjective. For the

injectivity let $\beta_1 * \overline{\lambda}_1 = \beta_2 * \overline{\lambda}_2$ with $\beta_1 = (p_1, q_1)$, $\beta_2 = (p_2, q_2)$ in B

and λ_1 , λ_2 in $U \cap O_K$. Then $p_1 \lambda_1^2 = \mu^4 p_2 \lambda_2^2$ and $q_1 \lambda_1^3 = \mu^6 q_2 \lambda_2^3$ for some

$\mu \in K^*$ (see 1.1) . We clearly have $\mu \in U$. Since β_1 , β_2 are basic

solutions, we conclude $v_\pi(\lambda_1) = v_\pi(\lambda_2) + 2 v_\pi(\mu)$ for all $\pi \in S$. Hence

$\delta := \mu^2 \lambda_2 / \lambda_1$ is in O_K^* and $p_1 = \delta^2 p_2$, $q_1 = \delta^3 q_2$. Since β_1 , β_2 are

in B we have $\delta = 1$ and $\beta_1 = \beta_2$, $\overline{\lambda}_1 = \overline{\lambda}_2$ as desired. The Theorem is

proved. ∎

(2.5) <u>Remarks.</u> (1) We already mentioned that B is a finite set.

On the other hand U/U^2 is also finite with $|U/U^2| = 2^{|S|+1+r_K}$, where

$r_K \in \{0,1\}$ is the rank of O_K^* . Hence $E(S)$ is finite with

$$|E(S)| = |B_S| \cdot 2^{|S|+1+r_K} \quad .$$

Note that $|S| \geq 2$ since S contains the prime divisors of 2 and 3 .

The finiteness of $E(S)$ holds also in case of an arbitrary number field K

with O_K not necessarily a principal ideal domain, due to the theorem of

Shafarevich (see [Lan 1; Chap. 2, Theorem 1]).

(2) Fix $\beta \in B$ and let $\lambda_1, \lambda_2 \in U$ such that $\overline{\lambda}_1 \neq \overline{\lambda}_2$. Then the

curves $\beta * \overline{\lambda}_1$ and $\beta * \overline{\lambda}_2$ are <u>$\lambda_1 \lambda_2^{-1}$- twists</u> of each other, i.e. they are

non-isomorphic over K and become isomorphic over $K(\sqrt{\lambda_1/\lambda_2})$.

Next we give an answer to the second of our above two questions and show how to get global minimal equations for the curves $\beta * \overline{\lambda}$. Let $v, w \in O_K$. Consider the equations

$$v = (a_1^2 + 4a_2)^2 - 24(a_1 a_3 + 2a_4)$$

$$w = -(a_1^2 + 4a_2)^3 + 36(a_1^2 + 4a_2)(a_1 a_3 + 2a_4) - 216(a_3^2 + 4a_6) \qquad .$$

Suppose that for some choice of a_1, a_2, a_3 with

$$a_1, \ a_3 \in \{\alpha_1 \omega_1 + \alpha_2 \omega_2 \mid \alpha_1, \alpha_2 = 0 \text{ or } 1\}$$

$$a_2 \in \{\alpha_1 \omega_1 + \alpha_2 \omega_2 \mid \alpha_1, \alpha_2 = -1, 0 \text{ or } 1\}$$

this pair of equations has a solution $(a_4, a_6) \in O_K \times O_K$. Then the numbers a_1, a_2, a_3, a_4, a_6 determine an integral Weierstrass equation Γ over K satisfying $v = c_4(\Gamma)$, $w = c_6(\Gamma)$. Different equations Γ arising from v, w in this way define the same elliptic curve over K , the curve given by the equation $\Gamma^{(1/6)} : y^2 = x^3 - 27vx - 54w$. If such Γ does exist we denote it by $\Gamma_{v,w}$. Note that we have the following <u>necessary condition for the existence of</u> $\Gamma_{v,w}$: If π is a prime divisor of 3 with $\pi | w$, then $\pi^3 | w$.

<u>(2.6) Lemma.</u> Let $v, w \in O_K$. Let E be an elliptic curve over K . The following conditions are equivalent.

(i) E has an integral Weierstrass equation Γ over K such that $v = c_4(\Gamma)$, $w = c_6(\Gamma)$.

(ii) E has a Weierstrass equation of type $\Gamma_{u^4 v, u^6 w}$ for any $u \in O_K^*$.

$\underline{\text{Proof.}}$ (i) implies (ii): Suppose that E has an equation Γ :
$y^2 + a_1xy + a_3y = x^3 + a_2x^2 + a_4x + a_6$ with a_i in O_K for all i and
$v = c_4(\Gamma)$, $w = c_6(\Gamma)$. Let $u \in O_K^*$. Choose $r,s,t \in O_K$ successively subject
to the following conditions:

$$ua_1 = a_1' + 2s$$
$$u^2a_2 = a_2' - sa_1' + 3r - s$$
$$u^3a_3 = a_3' + ra_1' + 2t$$

such that a_1', $a_2' \in \{\alpha_1\omega_1 + \alpha_2\omega_2 \,|\, \alpha_1,\alpha_2 = 0 \text{ or } 1\}$, $a_2' \in \{\alpha_1\omega_1 + \alpha_2\omega_2 \,|\, \alpha_1,\alpha_2 = -1, 0 \text{ or } 1\}$. Now we make the coordinate change $x' = u^2x + r$,
$y' = u^3y + u^2sx + t$. One obtains an equation Γ' for E

$$\Gamma' : \quad y'^2 + a_1'x'y' + a_3'y' = x'^3 + a_2'x'^2 + a_4'x' + a_6'$$

with $(a_4',a_6') \in O_K \times O_K$ and $u^4v = u^4c_4(\Gamma) = c_4(\Gamma')$, $u^6w = u^6c_6(\Gamma) = c_6(\Gamma')$
(see 1.1). Hence $\Gamma' = \Gamma_{u^4v,u^6w}$ and E has an equation of the desired
type.

(ii) implies (i) by construction of Γ_{u^4v,u^6w} putting $u = 1$. The
Lemma is proved. $\blacksquare$

Now we consider the following canonical set of representatives for
U/U^2 :

$$\mathbb{R} = \{\zeta^\mu \eta^\nu \pi_1^{\alpha_1} \cdots \pi_n^{\alpha_n} \,|\, \mu,\nu,\alpha_1,\ldots,\alpha_n = 0 \text{ or } 1\}$$

(for ζ, η see the notations). For $\lambda \in \mathbb{R}$ we simply write $\beta*\lambda$ instead
of $\beta*\overline{\lambda}$, $\beta \in B$.

(2.7) **Theorem.** Let $\beta = (p,q) \in B$, $\overline{\lambda} \in U/U^2$. If $\lambda \in \mathbb{R}$ then the equation

$$\Gamma : \quad y^2 = x^3 - 3\lambda^2 p x - 2\lambda^3 q$$

for $\beta * \lambda$ is minimal at all primes in O_K not dividing 2 or 3 . A global minimal Weierstrass equation over K for the curve $\beta * \lambda$ is given by some

$$\Gamma_{v,w} \quad ,$$

where $v,w \in O_K$ are of the form $v = u^{-4} 2^4 3^2 \lambda^2 p$, $w = u^{-6} 2^6 3^3 \lambda^3 q$ with some $u \in O_K$ divisible only by primes dividing 2 or 3 .

Proof. Let Γ' be a global minimal Weierstrass equation over K for $\beta * \lambda$. Then there exists $u \in K^*$ such that $u^{12} \Delta(\Gamma') = \Delta(\Gamma)$ and

$$2^4 3^2 \lambda^2 p = c_4(\Gamma) = u^4 c_4(\Gamma')$$
$$2^6 3^3 \lambda^3 q = c_6(\Gamma) = u^6 c_6(\Gamma') \quad .$$

Since $\Delta(\Gamma')$ divides $\Delta(\Gamma)$ in O_K we conclude that $u \in O_K$. Let $\pi \in O_K$ be a prime dividing u and suppose that π does not divide 2 or 3 . If $\pi \notin S$ then $v_\pi(\Delta(\Gamma)) = 0$ and Γ is clearly minimal at π . In particular $\pi \nmid u$. Let $\pi \in S$. Since $\beta = (p,q)$ is a basic solution and $\lambda \in \mathbb{R}$ we conclude from the above pair of equations that $v_\pi(u) = 0$ and Γ is minimal at π again. This proves the first part. Now $c_4(\Gamma')$, $c_6(\Gamma')$ are of the form $v = u^{-4} 2^4 3^2 \lambda^2 p$, $w = u^{-6} 2^6 3^3 \lambda^3 q$ resp. with $u \in O_K$ divisible only by primes dividing 2 or 3 . From Lemma 2.6 we conclude that $\Gamma_{v,w}$ exists and since $\Delta(\Gamma_{v,w}) = \Delta(\Gamma')$ the equation $\Gamma_{v,w}$ is a global minimal Weierstrass equation for $\beta*\lambda$. This proves the Theorem. ■

(2.8) __Remarks.__ (1) Theorem 2.7 involves an answer to the second of
our above two questions. Indeed, for any $\beta \in B$ and $\lambda \in R$ a global minimal Weierstrass equation for $\beta*\lambda$ over K may be constructed in the following
way. For any $v = u^{-4}2^4 3^2 \lambda^2 p$, $w = u^{-6}2^6 3^3 \lambda^3 q$ with $u \in O_K$ divisible only
by primes dividing 2 or 3 decide whether there exists $\Gamma_{v,w}$. In view
of Lemma 2.6 one has to consider u only up to associates, so in fact (v,w)
runs through a finite set. For at least one (v,w) the equation $\Gamma_{v,w}$
really does exist. Among all such equations $\Gamma_{v,w}$ you find one with discriminant dividing all the other discriminants. This is a desired global
minimal equation for $\beta*\lambda$.

(2) Let $\Gamma' = \Gamma_{v,w}$ be a global minimal Weierstrass equation for $\beta*\lambda$
with v,w as in the Theorem. Let π be a prime dividing 2 or 3 . Since
β is a basic solution and $\lambda \in R$, it follows from the pair of equations
in the proof of the Theorem that $2v_\pi(u) \leq 2v_\pi(2) + v_\pi(\lambda)$ in case $\pi|2$
and $2v_\pi(u) \leq v_\pi(3) + v_\pi(\lambda)$ in case $\pi|3$. So we obtain for $\Delta' = \Delta(\Gamma')$:

$$
(2.9) \quad v_\pi(\Delta') = \begin{cases}
6v_\pi(2)+6v_\pi(\lambda)+v_\pi(p^3-q^2)-12v_\pi(u) & \\
\quad \text{with } v_\pi(u) = 0,\ 1 \text{ or } 2 , & \text{if } \pi|2 . \\
\\
3v_\pi(3)+6v_\pi(\lambda)+v_\pi(p^3-q^2)-12v_\pi(u) & \\
\quad \text{with } v_\pi(u) = 0 \text{ or } 1 , & \text{if } \pi|3 . \\
\\
6v_\pi(\lambda)+v_\pi(p^3-q^2) , & \text{if } \pi \in S ,\ \pi \nmid 2,3 . \\
\\
0 & \text{if } \pi \notin S .
\end{cases}
$$

- 30 -

From Theorem 2.7 we can now derive necessary and sufficient conditions
for $\beta \in B$ and $\lambda \in \mathfrak{R}$ such that $\beta*\lambda$ has good reduction at a prime π
dividing 3 resp. 2 .

(2.10) <u>Corollary.</u> Let $\beta = (p,q) \in B$, $\lambda \in \mathfrak{R}$. Let π be a prime dividing 3 . Then the following conditions are equivalent.

   (i) $\beta*\lambda$ has good reduction at π .

   (ii) $6\nu_\pi(\lambda) + \nu_\pi(p^3 - q^2) = 12 - 3\nu_\pi(3)$, where $\nu_\pi(\lambda) \geq 2 - \nu_\pi(3)$,

       and a $\Gamma_{v,w}$ exists for some $v,w \in \mathcal{O}_K$ of the form

       $v = u^{-4}2^4 3^2 \lambda^2 p$, $w = u^{-6}2^6 3^3 \lambda^3 q$ with $u \in \mathcal{O}_K$ divisible only

       by primes dividing 2 or 3 and $\nu_\pi(u) = 1$.

<u>Proof.</u> Let Γ' be a global minimal equation for $\beta*\lambda$. By Theorem 2.7
we can assume that $\Gamma' = \Gamma_{v,w}$, where $v,w \in \mathcal{O}_K$ are of the form
$v = u^{-4}2^4 3^2 \lambda^2 p$, $w = u^{-6}2^6 3^3 \lambda^3 q$ with $u \in \mathcal{O}_K$ divisible only by primes
dividing 2 or 3 . (i) implies (ii) : Suppose that $\beta*\lambda$ has good re-
duction at π . Then by 2.9 we have $\nu_\pi(\Delta(\Gamma')) = 3\nu_\pi(3)+6\nu_\pi(\lambda)+\nu_\pi(p^3-q^2)$
$-12\nu_\pi(u) = 0$. We have $2\nu_\pi(u) \leq \nu_\pi(3)+\nu_\pi(\lambda)$, so that $\nu_\pi(u) \in \{0,1\}$.
We conclude that $\nu_\pi(u) = 1$ and $6\nu_\pi(\lambda)+\nu_\pi(p^3-q^2) = 12-\nu_\pi(3)$, as desired.

   (ii) implies (i) : Suppose that $\Gamma_{v,w}$ exists, where v,w have the
properties stated in (ii). Then $\Gamma_{v,w}$ is equivalent to $(\Gamma_{v,w})^{(1/6)}$:
$y^2 = x^3 - u^{-4}2^4 3^5 \lambda^2 px - u^{-6}2^7 3^6 q$, which is an equation for $\beta*\lambda$. We have

$$\Delta(\Gamma_{v,w}) = 2^6 3^3 \lambda^6 (p^3 - q^2)u^{-12} .$$

From the condition $6\nu_\pi(\lambda)+\nu_\pi(p^3-q^2) = 12-3\nu_\pi(3)$ we conclude that

$\nu_\pi(\Delta(\Gamma_{v,w})) = 0$. Hence $\beta * \lambda$ has good reduction at π and (i) holds. The Corollary is proved. ∎

We state without proof the following analogue Corollary for π dividing 2 .

(2.11) __Corollary.__ Let $\beta = (p,q) \in B$, $\lambda \in \mathbb{R}$. Let π be a prime dividing 2 . Then the following conditions are equivalent.

(i) $\beta * \lambda$ has good reduction at π .

(ii) For $\kappa = 1$ or $\kappa = 2$ we have $6\nu_\pi(\lambda) + \nu_\pi(p^3 - q^2) = 12\kappa - 6\nu_\pi(2)$, where $\nu_\pi(\lambda) \geq 2\kappa - 2\nu_\pi(2)$, and a $\Gamma_{v,w}$ exists for some $v,w \in O_K$ of the form $v = u^{-4}2^4 3^2 \lambda^2 p$, $w = u^{-6}2^6 3^3 \lambda^3 q$ with $u \in O_K$ divisible only by primes dividing 2 or 3 and $\nu_\pi(u) = \kappa$. ∎

As an illustration of the above concepts we consider the following

(2.12) __Special case.__ Assume that 2 is not split in O_K and that 3 is a prime in O_K , let $2 = \delta\pi^\sigma$ with $\pi \in O_K$ a prime, $\delta \in O_K^*$, $\sigma \in \{1,2\}$. We consider

$$S = \{\pi,3\} \quad ,$$

so $E(S)$ is the set of all elliptic curves over K with <u>conductor of the form $N = (\pi)^a(3)^b$</u> for some $a,b \in \mathbb{N}_0$. Now in this case B becomes a set of representatives of non-associated basic solutions of the equation

$$x^3 - y^2 = \xi \pi^{z_1} 3^{z_2} \quad .$$

Let $\beta = (p,q,\varepsilon,m,n) \in B$, $\lambda = \zeta^\mu \eta^\nu \pi^\alpha 3^\beta$ for some $\mu,\nu,\alpha,\beta \in \{o,1\}$. Then Theorem 2.4 states: The curve $\beta * \lambda$ with Weierstrass equation

$$y^2 = x^3 - 3\lambda^2 p x - 2\lambda^3 q$$

lies in $E(S)$ and conversely, each ellitpic curve in $E(S)$ arises uniquely in this way. In particular we have

$$|E(\dot{S})| = |B| \cdot 2^{3+r}K \quad ,$$

where $r_K \in \{0,1\}$ is the rank of O_K^* . Now let $(p,q,\varepsilon,m,n) \in B$ with $n = 3$. Suppose that for some $\lambda = \zeta^\mu \eta^\nu \pi^\alpha 3^\beta$ with $\mu,\nu,\alpha \in \{0,1\}$, $\beta = 1$ there exists $\Gamma_{v,w}$ for some $\gamma \in \{0,1,2\}$, where

$$v = \delta^4 \zeta^{2\mu} \eta^{2\nu} \pi^{4\sigma+2\alpha-4\gamma} p$$
$$w = \delta^6 \zeta^{3\mu} \eta^{3\nu} \pi^{6\sigma+3\alpha-6\gamma} q \quad .$$

Then Corollary 2.10 states: The equation

$$y^2 = x^3 - 3\lambda^2 p x - 2\lambda^3 q$$

defines an elliptic curve E over K with <u>conductor of the form $N = (\pi)^a$</u> for some $a \in \mathbb{N}_0$, where a global minimal Weierstrass equation for E over K is given by $\Gamma_{v,w}$ with v,w corresponding to the largest $\gamma \in \{0,1,2\}$ above. Here (p,q,ε,m,n) also satisfies the following necessary condition:

If a prime divisor τ of 3 divides q , then τ^3 divides q . Converse-ly, each elliptic curve over K with conductor $N = (\pi)^a$, $a \in \mathbb{N}_0$, arises uniquely in the above way. In particular we have

$$|E(\{\pi\})| \leq |B^{(3)}| \cdot 2^{2+r_K} \quad ,$$

where $B^{(3)} = \{(p,q,\varepsilon,m,n) \in B : n = 3\}$.

In the following chapter we study the exponential diophantine equation $x^3 - y^2 = \xi\pi_1^{z_1}\cdots\pi_n^{z_n}$ more closely.

Chapter 3

The diophantine equation $x^3 - y^2 = r$

Let S be a finite set of primes in 0_K . In chapter 2 we saw in case 0_K is a principal ideal domain how to construct the elliptic curves over K with good reduction outside S from a set B_S of representatives of non-associated basic solutions of the exponential diophantine equation

$$x^3 - y^2 = \xi \, \pi_1^{z_1} \cdots \pi_n^{z_n} \quad ,$$

where $S = \{\pi_1, \ldots, \pi_n\}$ (see Theorem 2.4). For that we had to assume that every prime divisor of 2 or 3 is contained in S . Solutions in B_S that give rise to elliptic curves with good reduction outside $S \smallsetminus S_1$, where S_1 is some set of prime divisors of 2 or 3 in 0_K , satisfy certain additional conditions (see Corollaries 2.10 and 2.11, see also 2.12) and hence for the construction of these curves the above equation need not be solved completely. In the present chapter we deal now with the construction of the set B_S .

Throughout this chapter we keep the above set S fixed. We assume that 0_K is a principal ideal domain, K quadratic or $K = \mathbb{Q}$. If A is a ring then for $r \in A$ we denote the equation

$$x^3 - y^2 = r$$

by Γ_r . For any ring extension R of A we let

$$\Gamma_r(R)$$

be the set of solutions (x,y) of Γ_r with x,y in R . An important choice for us will be $A = 0_K$ and $R = 0_K[S^{-1}] : = 0_K[\pi_1^{-1},\ldots,\pi_n^{-1}]$.

(3.1) <u>Remark.</u> Let R be a ring with $0_K \subseteq R \subseteq K$, $r \in 0_K$. Let π be a prime element in 0_K . Write $(s,t) \in \Gamma_r(R)$ in the form $s = e/f$, $t = g/h$ with $e,f,g,h \in 0_K$, $\gcd(e,f) = \gcd(g,h) = 1$, $f = \pi^{\alpha_1}d_1$, $g = \pi^{\alpha_2}d_2$ for some $d_1,d_2 \in 0_K$ with $\gcd(\pi,d_1) = \gcd(\pi,d_2) = 1$, $\alpha_i \geq 0$, $i \in \{1,2\}$. Then

$$\alpha_1 = 2\alpha \ , \ \alpha_2 = 3\alpha$$

for some $\alpha \geq 0$. Indeed, let $\alpha \geq 0$ be minimal such that $\alpha_1 \leq 2\alpha$, $\alpha_2 \leq 3\alpha$. Then equality holds. For, if $\alpha_1 = 0$ then $\alpha_2 = 0$ and so we can assume $\alpha_1 > 0$. We have

$$(e\pi^{2\alpha-\alpha_1})^3 d_2^2 - (g\pi^{3\alpha-\alpha_2})^2 d_1^3 = r\pi^{6\alpha}d_1^3 d_2^2 \qquad .$$

Suppose $\alpha_1 < 2\alpha$. Then we successively have $3\alpha - \alpha_2 \geq 2$, $2\alpha - \alpha_1 \geq 2$, $3\alpha - \alpha_2 \geq 3$. This is a contradiction to the minimality of α . Thus $\alpha_1 < 2\alpha$ is impossible. If $\alpha_2 < 3\alpha$, then $\alpha_1 < 2\alpha$ and hence $\alpha_2 < 3\alpha$ is also impossible. Thus we obtain $\alpha_1 = 2\alpha$, $\alpha_2 = 3\alpha$, as claimed.

In the following Lemma we show that the construction of a set B_S of non-associated basic solutions is equivalent to the study of a certain finite number of equations of type Γ_r . (For the meaning of W_K , ζ, η we refer to the notations.)

(3.2) __Lemma.__ Let $R = O_K[S^{-1}]$. There is a bijection

$$\Xi : \ B_S \ \rightarrow \ \bigcup_{r \in I} \Gamma_r(R) \qquad ,$$

where I denotes the finite set of all numbers $\zeta^{j_1}\eta^{j_2}\pi_1^{k_1}\cdots\pi_n^{k_n}$ with $0 \leq j_1 \leq |W_K/W_K^6|$, $0 \leq j_2, k_1, \ldots, k_n \leq 5$.

__Proof.__ Let $\beta = (p, q, \varepsilon, m , \ldots, m_n)$ be any basic solution. Write $\varepsilon = \zeta^{6+j_1}\eta^{6a+j_2}$, $m_\nu = 6a_\nu + k_\nu$ such that $a, a_1, \ldots, a_n \in \mathbb{N}_0$ and $0 \leq j_1 \leq |W_K/W_K^6|$, $0 \leq j_2, k_1, \ldots, k_n \leq 5$. Define

$$\Xi'(\beta) = (\frac{p}{\zeta^2\eta^{2a}\pi_1^{2a_1}\cdots\pi_n^{2a_n}} \ , \ \frac{q}{\zeta^3\eta^{3a}\pi_1^{3a_1}\cdots\pi_n^{3a_n}}) \qquad .$$

Then $\Xi'(\beta) \in \Gamma_r(R)$ for $r = \zeta^{j_1}\eta^{j_2}\pi_1^{k_1}\cdots\pi_n^{k_n}$. Ξ' is surjective. Indeed, let $(s,t) \in \Gamma_r(R)$ for some $r = \zeta^{j_1}\eta^{j_2}\pi_1^{k_1}\cdots\pi_n^{k_n} \in I$. Then by Remark 3.1 we can write $s = \dfrac{s_1}{\pi_1^{2a_1}\cdots\pi_n^{2a_n}}$, $t = \dfrac{t_1}{\pi_1^{3a_1}\cdots\pi_n^{3a_n}}$, where $s_1, t_1 \in O_K$ such that π_ν does not divide s_1 and t_1 , if $a_\nu > 0$, $1 \leq \nu \leq n$. Put $\beta = (s_1, t_1, \zeta^{j_1}\eta^{j_2}, 6a_1 + k_1, \ldots, 6a_n + k_n)$. Then β is clearly a basic solution and $\Xi'(\beta) = (s,t)$. Now it is easy to see that two basic solutions have the same image under Ξ' if and only if they are associated. Thus Ξ' induces a bijection $\Xi : B_S \rightarrow \bigcup_{r \in I} \Gamma_r(R)$, as claimed. $\blacksquare$

We point out that each equation Γ_r , $r \in I$, defines itself an elliptic curve over K in $E(S)$. Thus by Lemma 3.2 (together with Theorem 2.4) the set $E(S)$ is completely known if one knows for a certain finite subset of curves in $E(S)$ all points in $O_K[S^{-1}]$ with regard to a special Weierstrass equation for each of these curves.

We are now going to study equations of type Γ_r , $r \in O_K$, more closely (O_K principal ideal domain, K quadratic or $K = \mathbb{Q}$). The most important result is given in the following Theorem.

(3.3) Theorem. Let $R = O_K[T^{-1}]$, where T is any finite set of pairwise non-associated prime elements in O_K . Let $r \in O_K$. Then

$$\Gamma_r(R)$$

is finite.

Putting $T = S$, Lemma 3.2 and Theorem 2.4 yield:

(3.4) Corollary. B_S is finite. ∎

(3.5) Corollary. $E(S)$ is finite. ∎

Theorem 3.3 was first proven by Mahler [Mah] in a non-effective form. Mahler's result, valid for any curve of genus 1 over K , is an extension of Siegel's famous theorem (see [Si]) on integer points on curves of genus 1 . An effective proof of the theorem was given (in case $K = \mathbb{Q}$) by Coates [Coa,III], by means of Baker's methods. Coates' result is as follows. Let $T = \{p_1,\ldots,p_m\}$, $m \in \mathbb{N}_0$. Put $(s,t) \in \Gamma_r(R)$, $R = \mathbb{Z}[T^{-1}]$, in the form $s = \dfrac{s_1}{p_1^{2a_1}\cdots p_m^{2a_m}}$, $t = \dfrac{t_1}{p_1^{3a_1}\cdots p_m^{3a_m}}$ with $s_1, t_1 \in \mathbb{Z}$, $a_\mu \in \mathbb{N}_0$, where p_μ does not divide s_1 and t_1 , if $a_\mu > 0$, $1 \leq \mu \leq m$. Let d be the greatest divisor of $r \in \mathbb{Z}$, comprised solely of powers of $p_1,\ldots,p_m$. Put $r' = r/d$, $P = \max\{p_\mu \mid 1 \leq \mu \leq m\}$, $P = 2$ in case $m = 0$. Then

$$\max\{|s_1|,|t_1|\} < \exp\{2^{10^7(m+1)}P^{10^9(m+1)^3}|r'|^{10^6(m+1)^2}\} \quad .$$

Clearly, this estimation involves an algorithmical method for finding all $(s,t) \in \Gamma_r(R)$. Indeed, for each pair $(s_1,t_1) \in \mathbb{Z}\times\mathbb{Z}$ satisfying the above inequality one has to check whether $s_1^3 - t_1^2$ is of the form $rp_1^{6a_1}\cdots p_m^{6a_m}$ for some $a_\mu \in \mathbb{N}_0$, $1 \leq \mu \leq m$. This method however is impracticable for an explicit determination of $\Gamma_r(R)$, because the above upper bound is so large that it is far beyond the range of any computer. In the following we now describe how to proceed in practice.

There are the following two ways factorizing the equation $x^3 - y^2 = r$, $r \in O_K$, in a finite extension field L of K :

$$x^3 = (y + \sqrt{-r})(y - \sqrt{-r}) \quad , \text{ where } L = K(\sqrt{-r}) ,$$
$$y^2 = (x - \sqrt[3]{r})(x - \rho\cdot\sqrt[3]{r})(x - \rho^2\cdot\sqrt[3]{r}) \quad , \text{ where } L = K(\sqrt[3]{r},\rho) ,$$

ρ a primitive cube root of unity. Considering the arithmetical properties of the field L one can explicitly construct a finite set $\mathcal{S}_{r,L}$ of diophantine equations over K such that the solutions in $\Gamma_r(R)$, $R = 0_K[T^{-1}]$, can be parametrized by the solutions over R of the equations in $\mathcal{S}_{r,L}$. In other words, there is a surjective map

$$(3.6) \qquad \Pi_{r,L} : \bigcup_{F \in \mathcal{S}_{r,L}} V(F) \to \Gamma_r(R) \qquad ,$$

where $V(F)$ denotes the set of solutions of $F \in \mathcal{S}_{r,L}$ over R (the union is taken to be disjoint). In this way the determination of $\Gamma_r(R)$ is reduced to the determination of each $V(F)$, $F \in \mathcal{S}_{r,L}$. More precisely, in case $L = K(\sqrt{-r})$ each F in $\mathcal{S}_{r,L}$ is of type

$$f(X,Y) = s \qquad ,$$

where f is a binary cubic form with coefficients in $\Delta^{-1}0_K$, $s \in 0_K$, $\Delta \in 0_K$ a constant, depending only on L . In case $L = K(\sqrt[3]{r},\rho)$ each F in $\mathcal{S}_{r,L}$ is of the same type, but f is now a binary quartic form. There is a large number of results concerning diophantine equations of type $f(X,Y) = s$ with a binary cubic or quartic form f , $s \in 0_K$. For each equation in $\mathcal{S}_{r,L}$ one hopes "to see all solutions over R directly" or to be able to apply one of these known results in order to get all solutions over R explicitly. In the following we go into the details in case $L = K(\sqrt{-r})$. We first give the explicit definitions of the cubic forms f in $\mathcal{S}_r = \mathcal{S}_{r,L}$ and of the parametrization $\Pi_{\iota} = \Pi_{r,L}$ and then show, by means of the arithmetical properties of the field L , that $\mathcal{S}_r$ and Π_r have the desired properties.

We fix $r \in O_K$ and $T = \{\gamma_1, \ldots, \gamma_m\}$, where the γ_μ are pairwise nonassociated prime elements in O_K , and put $R = O_K[T^{-1}]$. We write

$$r = s^2 t \quad ,$$

with $s, t \in O_K$ such that t is squarefree in O_K . We assume that $t \neq -1$, the case $t = -1$ will be treated afterwards. Let L be the quadratic extension

$$L = K(\Theta) \quad , \quad \Theta^2 = -t \quad ,$$

of K . In $O_L[X,Y]$ the equation Γ_r becomes

$$x^3 = (y + s\Theta)(y - s\Theta) \quad .$$

Since O_K is a principal ideal domain, O_L is free of rank 2 over O_K . Let $\{\beta_1, \beta_2\}$ with

$$\beta_1 = a + c\Theta \quad , \quad \beta_2 = b + d\Theta \quad , \quad a,b,c,d \in K$$

be a basis of O_L over O_K . Put

$$M = \begin{pmatrix} a & b \\ c & d \end{pmatrix} \quad , \quad M = \Delta^{-1} M' \quad , \quad M' = \begin{pmatrix} a' & b' \\ c' & d' \end{pmatrix}$$

with $\Delta \in O_K$, $M' \in M_2(O_K)$. Now to any $\sigma + \tau\Theta \in L$, $\sigma, \tau \in K$ we associate binary cubic forms $g_{\sigma,\tau}$ $f_{\sigma,\tau}$ over K in the following way:

(3.7)
$$\begin{array}{l} g_{\sigma,\tau}(X,Y) = \sigma X^3 + 3\tau t X^2 Y + 3\sigma t XY^2 + \tau t^2 Y^3 \\ f_{\sigma,\tau}(X,Y) = \tau X^3 + 3\sigma X^2 Y + 3\tau t XY^2 + \sigma t Y^3 \end{array}$$

One immediately checks that in $L[X,Y] = K[X,Y] \oplus K[X,Y]\Theta$ the following identity holds:

(3.8)
$$(\sigma + \tau\Theta)(X + Y\Theta)^3 = g_{\sigma,\tau}(X,Y) + f_{\sigma,\tau}(X,Y)\Theta \quad .$$

Moreover, if for a binary cubic form $f(X,Y)$ over K and $\mathfrak{C} \in GL_2(K)$, $\mathfrak{C} = \begin{pmatrix} \alpha & \beta \\ \gamma & \delta \end{pmatrix}$, we denote the equivalent form $f(\alpha X + \beta Y, \gamma X + \delta Y)$ by $f^{\mathfrak{C}}(X,Y)$, then $g_{\sigma,\tau}$ and $f_{\sigma,\tau}$ satisfy the following <u>properties</u>:

(3.9)
$$g_{\sigma,\tau}(X,Y) = g^{\mathfrak{C}}_{\sigma,-\tau}(X,Y) \quad , \quad f_{\sigma,\tau}(X,Y) = f^{\mathfrak{C}}_{\sigma,-\tau}(X,Y) \quad ,$$

where $\mathfrak{C} = \begin{pmatrix} 1 & 0 \\ 0 & -1 \end{pmatrix}$.

(3.10) Let $\sigma + \tau\Theta = (\sigma_1 + \tau_1\Theta)(\sigma_2 + \tau_2\Theta)^3$, σ_i , $\tau_i \in K$, $i \in \{1,2\}$. Then
$$g_{\sigma,\tau}(X,Y) = g^{\mathfrak{C}}_{\sigma_1,\tau_1}(X,Y) \quad , \quad f_{\sigma,\tau}(X,Y) = f^{\mathfrak{C}}_{\sigma_1,\tau_1}(X,Y) \quad ,$$

where $\mathfrak{C} = \begin{pmatrix} \sigma_2 & \tau_2 t \\ \tau_2 & \sigma_2 \end{pmatrix}$.

Now we assume that $\sigma + \tau\Theta \in \mathcal{O}_L$. Then

$$g_{\sigma,\tau} \, , \, f_{\sigma,\tau} \in \Delta^{-1}\mathcal{O}_K[X,Y] \quad .$$

Indeed, we have $\sigma = \sigma'a + \tau'b = \Delta^{-1}(\sigma'a' + \tau'b')$ and $\tau = \sigma'c + \tau'd = \Delta^{-1}(\sigma'c' + \tau'd')$ for some $\sigma',\tau' \in \mathcal{O}_K$. Putting $\sigma_1 = \sigma'a' + \tau'b'$, $\tau_1 = \sigma'c' + \tau'd'$ we obtain $g_{\sigma,\tau} = \Delta^{-1}g_{\sigma_1,\tau_1}$ and $f_{\sigma,\tau} = \Delta^{-1}f_{\sigma_1,\tau_1}$ with g_{σ_1,τ_1} , $f_{\sigma_1,\tau_1} \in \mathcal{O}_K[X,Y]$, as claimed. From property 3.9 we conclude that the forms $f_{\sigma,\tau}$ and $f_{\sigma,-\tau}$ in $\Delta^{-1}\mathcal{O}_K[X,Y]$ are equivalent under

$GL_2(O_K)$. Moreover, if $\sigma + \tau\Theta$ and $\sigma_1 + \tau_1\Theta$ are in O_L^* such that $\sigma + \tau\Theta = (\sigma_1 + \tau_1\Theta)(\sigma_2 + \tau_2\Theta)^3$, then from property 3.10 we conclude that the forms $f_{\sigma,\tau}$ and f_{σ_1,τ_1} in $\Delta^{-1}O_K[X,Y]$ are equivalent under $GL_2(O_K[\Delta^{-1}])$. Indeed, for $\mathfrak{C}$ in property 3.10 we have $\mathfrak{C} \in M_2(O_K[\Delta^{-1}])$ and $\det\mathfrak{C} = \sigma^2 - \tau^2 t \in O_K^*$. For $\sigma + \tau\Theta \in O_L^*$ such that the norm $\sigma^2 - \tau^2 t$ in O_K^* is a cube in O_K^* , say $\delta_{\sigma,\tau}^3 = \sigma^2 - \tau^2 t$, $\delta_{\sigma,\tau} \in O_K^*$, we define a binary quadratic form $h_{\sigma,\tau} \in O_K[X,Y]$ in the following way:

(3.11)
$$\boxed{\; h_{\sigma,\tau}(X,Y) = \delta_{\sigma,\tau}X^2 - \delta_{\sigma,\tau}tY^2 \;}$$

From equation 3.8 we conclude that the forms $f_{\sigma,\tau}$, $g_{\sigma,\tau}$, $h_{\sigma,\tau}$ satisfy the following relation:

(3.12)
$$h_{\sigma,\tau}^3 - g_{\sigma,\tau}^2 = tf_{\sigma,\tau}^2$$

Now we can specify our set $\mathcal{S}_r$ and the parametrization Π_r occuring in 3.6 . Let

$$K_r$$

be a set of pairs $(\sigma,\tau) \in K \times K$ with $\sigma + \tau\Theta \in O_L^*$ satisfying the following property: The $\sigma + \tau\Theta$ and their conjugates (over K) $\sigma - \tau\Theta$ form a set of representatives in O_L^* modulo O_L^{*3}. K_r is clearly finite with

$$|K_r| = \frac{1}{2} (|O_L^*/O_L^{*3}| + 1)$$

From the above we conclude that for any $\sigma_1 + \tau_1\Theta \in O_L^*$ there exists (σ,τ) in K_r and $\mathfrak{C} \in GL_2(O_K[\Delta^{-1}])$ such that $f_{\sigma_1,\tau_1} = f_{\sigma,\tau}^{\mathfrak{C}}$, $g_{\sigma_1,\tau_1} = g_{\sigma,\tau}^{\mathfrak{C}}$.

More precisely, we can assume that $\mathcal{C}$ satisfies the following property:
If $u + v\Theta \in O_L$ and $\mathcal{C}\begin{pmatrix} u \\ v \end{pmatrix} = \begin{pmatrix} u' \\ v' \end{pmatrix}$, then $u' + v'\Theta \in O_L$. We put

$$\mathcal{S}_r = \{f_{\sigma,\tau}(X,Y) = s \mid (\sigma,\tau) \in K_r\} \quad ,$$

where $r = s^2 t$. For an equation $f_{\sigma,\tau}(X,Y) = s$ in $\mathcal{S}_r$ we define

$$V(f_{\sigma,\tau}) = \{(u,v) \in \Delta^{-1}R \times \Delta^{-1}R \mid f_{\sigma,\tau}(u,v) = s\} \quad .$$

In order to define the parametrization Π_r we make the following assumption.

$\underline{(3.13) \text{ Assumption.}}$ (i) $K \neq \mathbb{Q}(\sqrt{-3})$. (ii) K_r can be chosen such that for each $(\sigma,\tau) \in K_r$ the norm $\sigma^2 - \tau^2 t \in O_K^*$ is a cube in O_K^* . These assumptions are always satisfied , for example, if $K = \mathbb{Q}$ or K is imaginary quadratic, $K \neq \mathbb{Q}(\sqrt{-3})$. Note that if these assumptions are satisfied, then for each $(\sigma,\tau) \in K_r$ the number $\delta_{\sigma,\tau} \in O_K^*$ with $\sigma^2 - \tau^2 t = \delta_{\sigma,\tau}^3$ and hence the form $h_{\sigma,\tau}$ are uniquely determined by (σ,τ) .

For $(u,v) \in V(f_{\sigma,\tau})$, $(\sigma,\tau) \in K_r$, we put

$$\Pi_r(u,v) = (h_{\sigma,\tau}(u,v), g_{\sigma,\tau}(u,v)) \quad .$$

In view of relation 3.12 we obviously have defined in this way a map

$$\Pi_r : \bigcup_{(\sigma,\tau)\in K_r} V(f_{\sigma,\tau}) \to \Gamma_r(R[\Delta^{-1}]) \quad .$$

Our main result is now the following Theorem.

(3.14) Theorem. $\Gamma_r(R) \subseteq \mathrm{im}(\Pi_r)$.

For the proof of the theorem we need some further assumptions concerning

the field L .

(3.15) Assumption. (i) The class number of L is prime to 3 .

(ii) Each prime ideal $\mathfrak{p}$ in 0_L dividing $2r\gamma_1 \cdots \gamma_m$ is fixed under

the conjugation map.

The Theorem can also be proved without the assumptions 3.13 and 3.15 . Then

however, additional triples (f,g,h) of forms are necessary resp. the forms

$f_{\sigma,\tau}, g_{\sigma,\tau}, h_{\sigma,\tau}$ become more complicated.

Proof of the Theorem. Let $(p,q) \in \Gamma_r(R)$. We have to show that

$(p,q) = \Pi_r(u,v)$ for some $u,v \in \Delta^{-1}R$ with $f_{\sigma,\tau}(u,v) = s$ for some

$(\sigma,\tau) \in K_r$. We have $p^3 - q^2 = s^2 t$, $p = \dfrac{p_1}{\gamma_1^{2a_1} \cdots \gamma_m^{2a_m}}$, $\dfrac{q_1}{\gamma_1^{3a_1} \cdots \gamma_m^{3a_m}} = q$,

where $a_\mu \in \mathbf{N}_0$, $p_1, q_1 \in 0_K$ such that γ_μ does not divide p_1 and q_1 ,

if $a_\mu > 0$, $1 \le \mu \le m$. We have

$$p_1^3 = (q_1 + \gamma_1^{3a_1} \cdots \gamma_m^{3a_m} s\Theta)(q_1 - \gamma_1^{3a_1} \cdots \gamma_m^{3a_m} s\Theta) \qquad .$$

Let $\mathfrak{p}$ be a prime ideal in 0_L dividing both factors on the right.

Subtracting the factors from each other we obtain that $\mathfrak{p}$ divides $2r\gamma_1\cdots\gamma_m$. Thus we conclude

$$(3.16) \qquad (q_1 + \gamma_1^{3a_1}\cdots\gamma_m^{3a_m} s\Theta) = \mathfrak{p}_1^{e_1}\cdots\mathfrak{p}_k^{e_k} I^3$$

for some ideal I in 0_L and prime divisors $\mathfrak{p}_i$ of $2r\gamma_1\cdots\gamma_m$, $0 \le e_i \le 2$, $1 \le i \le k$. By assumption 3.15 $\mathfrak{p}_i = \mathfrak{p}_i'$, $1 \le i \le k$, where $\mathfrak{p}_i'$ is the conjugate ideal of $\mathfrak{p}_i$ under $\Theta \to -\Theta$. We conclude

$$(q_1 - \gamma_1^{3a_1}\cdots\gamma_m^{3a_m} s\Theta) = \mathfrak{p}_1^{e_1}\cdots\mathfrak{p}_k^{e_k} I'^3$$

and hence

$$(p)^3 = \mathfrak{p}_1^{2e_1}\cdots\mathfrak{p}_k^{2e_k} (I\cdot I')^3 \qquad .$$

Thus each $2e_i$ is divisible by 3 and hence $e_i = 0$, $1 \le i \le k$. By equation 3.16 I^3 is now a principal ideal. By our assumption in 3.15 on the class number of L , I itself must be a principal ideal. Let $I = (u' + v'\Theta)$, $u' + v'\Theta \in 0_L$, $u',v' \in \Delta^{-1}0_K$. Then from 3.16 we conclude that

$$q_1 + \gamma_1^{3a_1}\cdots\gamma_m^{3a_m} s\Theta = (\sigma_1 + \tau_1\Theta)(u' + v'\Theta)^3$$

for some $\sigma_1 + \tau_1\Theta \in 0_L^*$, $\sigma_1,\tau_1 \in \Delta^{-1}0_K$. Now we consider the forms f_{σ_1,τ_1} and g_{σ_1,τ_1} . From equation 3.8 we conclude

$$q_1 = g_{\sigma_1,\tau_1}(u',v')$$
$$\gamma_1^{3a_1}\cdots\gamma_m^{3a_m} s = f_{\sigma_1,\tau_1}(u',v')$$

For f_{σ_1,τ_1} and g_{σ_1,τ_1} there exists $\mathfrak{C} \in GL_2(O_K[\Delta^{-1}])$ with $f^{\mathfrak{C}}_{\sigma_1,\tau_1} = f_{\sigma,\tau}$ and $g^{\mathfrak{C}}_{\sigma_1,\tau_1} = g_{\sigma,\tau}$ for some $(\sigma,\tau) \in K_r$, where we can assume for $\mathfrak{C}$ that $(1,\Theta)\mathfrak{C}\begin{pmatrix} u' \\ v' \end{pmatrix} \in O_L$. Hence

$$q_1 = g_{\sigma,\tau}(u_1,v_1)$$

$$\gamma_1^{3a_1} \cdots \gamma_m^{3a_m} s = f_{\sigma,\tau}(u_1,v_1) \quad ,$$

for some $u_1,v_1 \in \Delta^{-1}O_K$. From equation 3.12 we conclude that

$$h_{\sigma,\tau}(u_1,v_1)^3 = q_1^2 - \gamma_1^{6a_1} \cdots \gamma_m^{6a_m} s^2 t = p_1^3$$

and hence $p_1 = h_{\sigma,\tau}(u_1,v_1)$ since $K \neq \mathbb{Q}(\sqrt{-3})$. Thus we obtain

$$p = h_{\sigma,\tau}(u,v) \quad , \quad q = g_{\sigma,\tau}(u,v) \quad ,$$

where $u = \dfrac{u_1}{\gamma_1^{2a_1} \cdots \gamma_m^{2a_m}}$, $v = \dfrac{v_1}{\gamma_1^{3a_1} \cdots \gamma_m^{3a_m}}$ satisfy $f_{\sigma,\tau}(u,v) = s$, $u,v \in \Delta^{-1}R$. Hence $\Pi_r(u,v) = (p,q)$, as desired. Thus the Theorem is proved. ∎

(3.17) Remark. Let K be imaginary quadratic or $K = \mathbb{Q}$ such that $O_L = O_K[\Theta]$. Then $\Delta = 1$ and $f_{\sigma,\tau}$, $(\sigma,\tau) \in K_r$, has coefficients in O_K . In [Las 5] it is proved that $\{f_{\sigma,\tau} \mid (\sigma,\tau) \in K_r\}$ is a full set of representatives for the equivalence classes of binary cubic forms over O_K with discriminant

$$D = \delta^2 2^2 3^3 t \quad , \quad \delta \in O_K^* \quad .$$

We finally have to treat the case $t = -1$, i.e $r = -s^2$. Let

$$\sigma_1, \ldots, \sigma_k$$

in 0_K precisely be the prime divisors of $\gamma_1 \cdots \gamma_m s$ (up to associates) . We consider a set

$$K_{-s^2}$$

of chains $(e_1, \ldots, e_k; e_1', \ldots, e_k')$ of non-zero integers satisfying the following properties: (i) $e_i + e_i' = 0$ or 3 , $1 \leq i \leq k$. (ii) Put $e = (e_1, \ldots, e_k)$, $e' = (e_1', \ldots, e_k')$. If then (e, e') and (f, f') are two such chains in K_{-s^2} , then $(e, e') \neq (f', f)$. For each $(e, e') \in K_{-s^2}$ we define binary cubic forms $f_{e,e'}$, $g_{e,e'}$ with coefficients in $2^{-1}0_K$ and a binary quadratic form $h_{e,e'}$ in $0_K[X,Y]$ in the following way:

$$
f_{e,e'}(X,Y) = \frac{\sigma_1^{e_1} \cdots \sigma_k^{e_k}}{2} X^3 - \frac{\sigma_1^{e_1'} \cdots \sigma_k^{e_k'}}{2} Y^3 \quad .
$$

(3.18)
$$
g_{e,e'}(X,Y) = \frac{\sigma_1^{e_1} \cdots \sigma_k^{e_k}}{2} X^3 - \frac{\sigma_1^{e_1'} \cdots \sigma_k^{e_k'}}{2} Y^3 \quad .
$$

$$
h_{e,e'}(X,Y) = \sigma_1^{\frac{e_1+e_1'}{3}} \cdots \sigma_1^{\frac{e_k+e_k'}{3}} XY \quad .
$$

Now we can specify our set K_{-s^2} and the parametrization Π_{-s^2} occuring in 3.6 . We put

$$\mathcal{S}_{-s^2} = \{ f_{e,e'}(X,Y) = s \mid (e, e') \in K_{-s^2} \} \quad .$$

For an equation $f_{e,e'}(X,Y) = s$ in $\mathcal{S}_{-s^2}$ we define

$$V(f_{e,e'}) = \{ (u, v) \in R \times R \mid f_{e,e'}(u, v) = s \} \quad .$$

For $(u,v) \in V(f_{e,e'})$, $(e,e') \in K_{-s^2}$, we put

$$\Pi_{-s^2}(u,v) = (h_{e,e'}(u,v), g_{e,e'}(u,v)) \quad .$$

In this way we obviously get a map

$$\Pi_{-s^2} : \bigcup_{(e,e') \in K_{-s^2}} V(f_{e,e'}) \to \Gamma_{-s^2}(R[\tfrac{1}{2}]) \quad .$$

Using the unique prime factor decomposition in O_K it is an exercise, left to the reader, to show that also in this case the following Theorem holds:

(3.19) Theorem. $\Gamma_{-s^2}(R) \subseteq im(\Pi_{-s^2})$. ∎

Once the set K_r is explicitly known, then by Theorem 3.14 resp. 3.19 the determination of $\Gamma_r(R)$ is reduced to the solution of a finite number of equations of Thue's type

$$f(X,Y) = s \quad ,$$

where the binary cubic form f has coefficients in $\Delta^{-1}O_K$, $r = s^2 t$, $X,Y \in \Delta^{-1}R$, $\Delta \in O_K$ a constant. Effectivity results for Thue's equation (over $\mathbb{Z}$), obtained by means of Baker's methods, may be found in [Coa,I] . These results however can in general not be used for an explicit solution. On the other hand Thue's equation has extensively been treated in several books, where, among all, we want to mention [Mo] and [Bo&Sha] . In any case one may hope that one of the methods, described in these books, explicitly yield all solutions over $\Delta^{-1}R$ for our equations. We finally want to point out that considering the equations $f(X,Y) = s$, instead of Γ_r , has a great advantage for experimental investigations. Indeed, let (p,q) be a solution in $\Gamma_r(R)$ with great absolute values of the numerators of p and q . Suppose that $u,v \in \Delta^{-1}R$ have the property $f(u,v) = s$ and $\Gamma_r(u,v) = (p,q)$. Then all numerical examples show that the absolute values of the numerators of u and v keep very small.

For the construction of a set K_r with the desired properties one needs to explicitly know a system of fundamental units in L and a generator for the group of roots of unity in L . To illustrate this we discuss some examples.

(3.20) <u>Example ([Las 2])</u>. Let $K = \mathbb{Q}(i)$. We put $T = \{1+i,3\}$ and for r we consider all values $r = -i^j(1+i)^k 3^l$ for $j \in \{0,1\}$, $0 \leq k,l \leq 5$. By Lemma 3.2 the determination of $\Gamma_r(R)$, $R = \mathcal{O}_K[T^{-1}]$, for that T and all those r precisely yields the set

$$B_S$$

for $S = \{1+i,3\}$. (Here we have chosen $\zeta = -i$.) We put $\mathcal{O} = \mathcal{O}_K = \mathbb{Z}[i]$. We write r in the form

$$r = s^2 t$$

with $s,t \in \mathcal{O}$, t squarefree. Clearly, t is uniquely determined up to plus or minus sign. Fixing a sign for t we conclude that

$$t \in \{-1,-i,-(1+i),-i(1+i),3,-3i,-3(1+i),-3i(1+i)\} \qquad .$$

(Choosing +3 instead of -3 has technical reasons.) Then

$$s = \begin{cases} (1+i)^{a_1}3^{a_2} & \text{if} \quad t \neq 3 \\ i(1+i)^{a_1}3^{a_2} & \text{if} \quad t = 3 \end{cases}$$

for some suitable $a_1,a_2 \in \{0,1,2\}$. Suppose that $(p,q) \in \Gamma_r(R)$ with
r such that $t = -i(1+i)$ or $-3i(1+i)$. Then $(-\bar{p},i\bar{q}) \in \Gamma_{r'}(R)$ with
r' such that $t = -(1+i)$ resp. $-3(1+i)$. Hence the cases of r for which
$t = -i(1+i)$ or $t = -3i(1+i)$ need not be discussed separately. Now we give
the set K_r for the remaining values of r .

Case t = -3i. Then $\Theta^2 = 3i$, $L = \mathbb{Q}(\Theta)$. We have $O_L = O[\Theta]$,
$O_L^* = <i>\times<1+i + \Theta>$. Hence

$$K_r = \{(1,0),(1+i,1)\}$$

obviously has the desired properties.

Case t = -(1+i). Then $\Theta^2 = 1+i$, $L = \mathbb{Q}(\Theta)$. We have $O_L = O[\Theta]$,
$O_L^* = <i>\times<1 + \Theta>$. Hence

$$K_r = \{(1,0),(1,1)\}$$

obviously has the desired properties.

Case t = -3(1+i). Then $\Theta^2 = 3(1+i)$, $L = \mathbb{Q}(\Theta)$. We have $O_L = O[\Theta]$,

$0_L^* = <i>\times<2+i + \Theta>$. Hence

$$K_r = \{(1,0),(2+i,1)\}$$

obviously has the desired properties.

Case $t = 3$. Then $\Theta^2 = -3$. For $L = \mathbb{Q}(i,\Theta)$, the field of 12th root of unity, we have $0_L = 0[\rho]$, $\rho = \frac{-1+\sqrt{-3}}{2}$, $0_L^* = <i\rho>\times<\eta>$, $\eta = i + (-1+i)\Theta$. The set K_r of all (σ,τ) such that $\sigma + \tau\Theta \in \{1,\eta,\rho^2,\rho^2\eta^{-1},\rho^2\eta\}$ has the desired properties. (Note that $\Delta = 2$ in this case). We have

$$K_r = \{(1,0),(\tfrac{1+i}{2},\tfrac{-1+i}{2}),(-\tfrac{1}{2},-\tfrac{1}{2}),(\tfrac{-2+i}{2},-\tfrac{i}{2}),(\tfrac{2+i}{2},-\tfrac{i}{2})\}$$

Case $t = -i$. Then $\Theta^2 = i$, $L = \mathbb{Q}(\Theta)$, the field of 8th root of unity. We have $0_L = 0[\Theta]$, $0_L^* = <\Theta> <1 + (1-i)\Theta>$. Hence

$$K_r = \{(1,0),(1,1-i)\}$$

obviously has the desired properties.

In each of the above cases assumption 3.13 is clearly satisfied. So in each case we have constructed the forms $f_{\sigma,\tau}$, $g_{\sigma,\tau}$, $h_{\sigma,\tau}$ over 0 as given in 3.7 resp. 3.11, $(\sigma,\tau) \in K_r$. Moreover, in case $t = -3i, -(1+i), -3(1+i)$, 3 assumption 3.15 is also satisfied. Hence Theorem 3.15 holds in these cases. Thus for the determination of $\Gamma_r(R)$ in these cases one is led to solve

$$f_{\sigma,\tau}(X,Y) = (1+i)^{a_1}3^{a_2} \quad \text{resp.} \quad i(1+i)^{a_1}3^{a_2} \quad ,$$

over $0[\tfrac{1}{1+i},\tfrac{1}{3}]$ for $(\sigma,\tau) \in K_r$ with suitable $a_1,a_2 \in \{0,1,2\}$. In case

$t = -i$ assumption 3.15 is not satisfied since 3 splits in 0_L ,
$(3) = (1-i+\Theta)(1-i-\Theta)$. In this case Theorem 3.14 holds with the following
set K_r' instead of K_r :

$$K_r' = K_r \cup \{(3-3i,3),(6,3-6i),(-6i,-3-6i)\} \quad .$$

Case $t = -1$. Then

$$K_r = \{(0,0;0,0),(0,1;0,2),(1,0;2,0),(1,1;2,2),(1,2;2,1)\}$$

has the desired properties.

We conclude our discussion on the connection between the equation Γ_r
and binary cubic forms. We now briefly want to mention quite another aspect
concerning the equation Γ_r , which can be used in order to obtain all
solutions in $\Gamma_r(R)$, $R = 0_K[T^{-1}]$, for certain $r \in 0_K$. We assume that K
is quadratic and $r \in \mathbb{Z}$, $r \neq 0$. Then the equation Γ_r defines an elliptic
E_r over $\mathbb{Q}$. For E_r we have a "trace map"

$$\sigma_r : E_r(K) \to E_r(\mathbb{Q})$$

given by $\sigma_r(Q) = Q + Q'$, where $Q \to Q'$ denotes the conjugation on E
arising from the conjugation in K . (For the addition on the set $E_r(K)$
of K-rational points on E_r we refer the reader to the first part of chapter 4.) Writing σ_r in terms of the Weierstrass equation Γ_r of E_r we
obtain in this way a map

$$\sigma_{r,R} : \Gamma_r(R) \to \Gamma_r(\mathbb{Q}) \cup \{\underline{0}\} \quad ,$$

where $\underline{0}$ denotes the unique point of Γ_r at infinity. The determination of $\Gamma_r(R)$ amounts now to the determination of all fibres $\sigma_{r,R}^{-1}(P)$, $P \in \Gamma_r(\mathbb{Q})$ $\cup \{\underline{0}\}$. In particular, if $E_r(\mathbb{Q})$ is a finite group, one has to consider only a finite number of such fibres. How the fibres can in many cases be determined is described in [Las 3] . A modification of the trace map also allows the determination of $\Gamma_r(R)$ for certain $r \in \mathcal{O}_K$, not necessarily in $\mathbb{Z}$.

Chapter 4

Isogeny Classes

In this chapter we study the isogeny classes of the elliptic curves
over K . Our main goal is to show how for the elliptic curves over K
with given conductor the division into isogeny classes can in many cases
explicitly be carried out. This corresponds to step (3) in the introduction.
We will also see how to visualize an isogeny class by a certain type of
graphs.

To begin with we recall some fundamental facts concerning the group
structure on an elliptic curve. For this we fix an elliptic curve E over
K (K arbitrary number field). E has a Weierstrass equation Γ of the
form

$$(4.1) \qquad\qquad \Gamma : \quad y^2 = x^3 + ax + b$$

with $a,b \in K$, and we even can assume $a,b \in \mathcal{O}_K$. Indeed, if Γ' is any
integral Weierstrass equation for E over K , then $\Gamma = (\Gamma')^{(1/6)}$ (see
1.5) has the desired property. We assume $\overline{K}$ to be embedded into $\mathbb{C}$ and
consider an intermediate field $K \subseteq L \subseteq \mathbb{C}$. Let

$$E(L)$$

be the vanishing set of the homogenized polynomial Γ in the projective space $\mathbb{P}^2(L)$. Here the coordinates $(x,y) = (x,y,1)$ correspond to the affine part of this set and $\underline{0} = (0,1,0)$ is the unique point of $E(L)$ at infinity. $E(\mathbb{C})$ carries the structure of a commutative algebraic group defined over K in the following way. $\underline{0}$ becomes the neutral element and for $P_1 \neq \underline{0}$, say $P_1 = (\xi_1,\eta_1)$ the inverse is given by $-P = (\xi_1,-\eta_1)$. Furthermore, for $P_1 \neq \underline{0}$, $P_2 \neq \underline{0}$, $P_2 = (\xi_2,\eta_2)$, $P_2 \neq P_1$ their sum $P_3 = (\xi_3,\eta_3)$ is given by

$$\xi_3 = \lambda^2 - \xi_1 - \xi_2$$
$$\eta_3 = - (\xi_3 - \xi_1) - \eta_1 \quad ,$$

where

$$\lambda = \begin{cases} \dfrac{\eta_1 - \eta_2}{\xi_1 - \xi_2} & , \quad \text{if} \quad P_1 \neq P_2 \ . \\[2mm] \dfrac{3\xi_1^2 + a}{2\eta_1} & , \quad \text{if} \quad P_1 = P_2 \ . \end{cases}$$

These formulas do indeed define the structure of a commutative algebraic group on $E(\mathbb{C})$ and it is clear that $E(L)$ becomes a subgroup. The above formulas arise from the following geometrical interpretation: Let $P_1, P_2, P_3 \in E(L)$. Then $P_1 + P_2 + P_3 = \underline{0}$ if and only if P_1, P_2, P_3 are on a line. If the equation

$$\Gamma' : \quad y^2 = x^3 + a'x + b'$$

with $a',b' \in K$ is also a Weierstrass equation for E over K , then there exists a transformation $x \to u^2 x$, $y \to u^3 y$ with $u \in K^*$, carrying Γ to Γ' (see chapter 1). From this we conclude that the group $E(L)$ is up to K-isomorphism uniquely determined by the equation Γ . More generally, if E' is an elliptic curve over K isomorphic to E over L , then the groups $E(L)$ and $E'(L)$ are isomorphic with an isomorphism defined over L , the isomorphism induced from the isomorphism from E to E' . Now in case $L = \mathbb{C}$ the converse of this last statement is true. Viewed as a complex analytic group, $E(\mathbb{C})$ is isomorphic to a torus

$$\mathbb{C}/\Lambda \quad ,$$

where Λ is a suitable lattice in $\mathbb{C}$ and an isomorphism from $\mathbb{C}/\Lambda$ to $E(\mathbb{C})$ is established by the Weierstrass $\wp_\Lambda$ and $\wp'_\Lambda$ functions. For all this see for example [Lan 1, Chapter 1 and 3] .

For a subset M of $E(\mathbb{C}) \subseteq \mathbb{P}^2(\mathbb{C})$ it is convenient to keep the following <u>notation</u>:

$$M^{\#} \quad , \quad M_2 \quad , \quad M_x \quad .$$

Here $M^{\#} = M - \{\underline{0}\}$, M_2 denotes the set of elements of order 2 in $M^{\#}$ and M_x denotes the set of x-coordinates of points in $M^{\#}$.

The absolute Galoisgroup G over K acts on $E(\overline{K})$ in a natural way. Indeed, G acts on $\mathbb{P}^2(\overline{K})$ in a natural way and $E(\overline{K})$ is invariant under this action. We denote this action by $P \to P^\sigma$, $P \in E(\overline{K})$, $\sigma \in G$. We have the following properties: $(P + Q)^\sigma = P^\sigma + Q^\sigma$ for $P,Q \in E(\overline{K})$, $\sigma \in G$, and hence $P \to P^\sigma$ defines an automorphism of the group $E(\overline{K})$. A subgroup U of $E(\overline{K})$ is invariant under G if and only if the set U_x is invariant under G . The set of fixed points of G on $E(\overline{K})$ is $E(K)$, the

<u>Mordell-Weil group</u> of E over K . The Mordell-Weil theorem states that the abelian group $E(K)$ is finitely generated and hence is of the form

$$E(K) \simeq \mathbb{Z}^r \times E(K)_{tors} \quad ,$$

where r is the rank of $E(K)$ and $E(K)_{tors}$ the torsion subgroup of $E(K)$.

For $n \in \mathbb{N}$ let $D_n(E)$ be the set of <u>n-division points</u> of $\underline{0}$ in $E(\mathbb{C})$:

$$D_n(E) = \{ P \in E(\mathbb{C}) \mid nP = \underline{0} \} \quad ,$$

where $nP = P + \dots + P$ (n times) . Representing $E(\mathbb{C})$ analytically as $\mathbb{C}/\Lambda$ for some suitable lattice Λ in $\mathbb{C}$, it is an easy exercise to verify that $D_n(E)$ is finite with

$$D_n(E) \simeq \mathbb{Z}/n\mathbb{Z} \times \mathbb{Z}/n\mathbb{Z} \quad .$$

Sometimes it is useful to express $D_n(E)$ in terms of coordinates: Let $\delta_n \in \{0,1\}$ be such that $n + 1 \equiv \delta_n \bmod 2$, let $s_n = \dfrac{n^2 - 3\delta_n - 1}{2}$. Then $D_n(E)$ can be written as the following disjoint union:

$$D_n(E) = \{ (\zeta_i, 0) \mid 1 \leq i \leq 3\delta_n \} \cup \{ (\xi_i, \pm\eta_i) \mid 1 \leq i \leq s_n \} \cup \{ \underline{0} \} \quad ,$$

where the first set on the right, if not empty (i.e. $\delta_n = 1$), equals $D_n(E)_2 = D_2(E)$, the ξ_i are pairwise different and $\eta_i \neq 0$, $1 \leq i \leq s_n$. We have

$$D_n(E) \subset E(\overline{K}) \qquad .$$

More precisely, one can explicitly construct a polynomial $T_n(E)$ in $O_K[x]$, even with coefficients in $\mathbb{Z}[a,b]$, such that the set of roots of $T_n(E)$ is precisely $D_n(E)_x$. $T_n(E)$ can recursively be defined in the following way. Let

$$f_1 = 1$$
$$f_2 = 2y$$
$$f_3 = 3x^4 + 6ax^2 + 12bx - a^2$$
$$f_4 = 4y(x^6 + 5ax^4 + 20bx^3 - 5a^2x^2 - 4abx - 8b^2 - a^3)$$

and for $m \geq 5$ let

$$f_m = \begin{cases} f_{k+2} \, f_k^3 - f_{k+1}^3 \, f_{k-1} & , \text{ if } m = 2k + 1 \ . \\[2ex] \dfrac{f_k(f_{k+2} \, f_{k-1}^2 - f_{k-2} \, f_{k+1}^2)}{2y} & , \text{ if } m = 2k \ . \end{cases}$$

Put

$$(4.2) \qquad T_n(E) = \begin{cases} f_n & , \text{ if } n \text{ odd} \ . \\[2ex] \frac{1}{2}y \, f_n & , \text{ if } n \text{ even} \ . \end{cases}$$

Then in fact $T_n(E)$ has the desired properties (see [Lan 2, Chapter II]): It is a polynomial in $\mathbb{Z}[a,b][x]$ such that the set of roots is precisely $\{\zeta_i \mid 1 \leq i \leq 3\delta_n\} \cup \{\xi_i \mid 1 \leq i \leq s_n\}$. In particular $T_n(E)$ has degree

$$\frac{n^2 + 3\delta_n - 1}{2}$$

with leading coefficient $\dfrac{n}{\delta_n + 1}$. The polynomial $T_n(E)$ is called the n-division polynomial of E over K . The field $K(D_n(E))$ obtained from K by adjoining the coordinates of all points in $D_n(E)$ is called the n-division field of E over K . Since, as one can easily see, $D_n(E)$ is invariant under the action of the Galoisgroup G on $E(\overline{K})$, we conclude that $K(D_n(E))$ is a finite Galois extension of K . Put

$$G_n(E) = \mathrm{Gal}(K(D_n(E)),K) \qquad .$$

Let $P_1,P_2 \in E(\mathbb{C})$ with $P_1 = (\xi_1,\eta_1)$, $P_2 = (\xi_2,\eta_2)$ be such that $D_n(E) = (\mathbb{Z}/n\mathbb{Z})P_1 \oplus (\mathbb{Z}/n\mathbb{Z})P_2$. Then $K(D_n(E)) = K(\xi_1,\xi_2,\eta_1,\eta_2)$. Moreover, to each $\sigma \in G_n(E)$ one can attach a matrix $\begin{pmatrix} a & b \\ c & d \end{pmatrix}$ determined by the equations

$$P_1^\sigma = aP_1 + bP_2 \quad , \quad P_1^\sigma = cP_1 + dP_2 \qquad .$$

This defines a faithful representation R_n of $G_n(E)$ in $GL_2(\mathbb{Z}/n\mathbb{Z})$. In particular, the degree of $K(D_n(E))$ over K is a divisor of $|GL_2(\mathbb{Z}/n\mathbb{Z})|$, which equals

$$\phi(n)\, n^3 \prod_{p \mid n} \left(1 - \frac{1}{p^2}\right) \qquad ,$$

where ϕ is the Euler ϕ-function. If P_1 can be chosen such that the subgroup $U_1 = \langle P_1 \rangle$ is invariant under G , then the representation R_n takes all its values in the subgroup of upper triangular matrices of $GL_2(\mathbb{Z}/n\mathbb{Z})$ and, if P_2 can be chosen such that $U_2 = \langle P_2 \rangle$ is also invariant, then all the values are diagonal matrices in $GL_2(\mathbb{Z}/n\mathbb{Z})$.

The above functions f_ν do not only give rise to the n-division points of $\underline{0}$ but to the n-division points of any point Q on $E(\mathbb{C})$. Indeed, let $Q \neq \underline{0}$, say $Q = (\xi,\chi)$. Let $P = (x,y)$ be a n-division point of Q on $E(\mathbb{C})$ for some $n \in \mathbf{N}$, so that

- 60 -

$$n\,P = Q \quad .$$

Then from the addition formulas we know that

$$\xi = g_n(x,y) \quad , \quad \chi = h_n(x,y)$$

for some rational functions $g_n = g_n(E)$, $h_n = h_n(E)$ with coefficients in K , both defined at $P = (x,y)$. Now g_n , h_n may explicitly be given in terms of the f_ν's in the following way. Define ϕ_n , ω_n $(n > 2)$ as follows:

$$\phi_n = x\, f_n^2 - f_{n-1}\, f_{n+1}$$
$$4y\omega_n = f_{n+2}\, f_{n-1}^2 - f_{n-2}\, f_{n+1}^2 \quad .$$

Then for $n > 2$ we have

$$(4.3) \qquad\qquad (g_n, h_n) = \left(\frac{\phi_n}{f_n^2}\ ,\ \frac{\omega_n}{f_n^3}\right)$$

(see [Lan 2,Chapter II]). Note that g_n can be written as a rational function only in the variable x .

Let U be a finite subgroup of order n in $E(\mathbb{C})$. Then $U \subseteq D_n(E) \subset E(\overline{K})$ and, since $D_n(E) \simeq \mathbb{Z}/n\mathbb{Z} \times \mathbb{Z}/n\mathbb{Z}$, there are precisely $\psi(n)$ such subgroups U , where

$$\psi(n) = n \prod_{p \mid n} \left(1 + \frac{1}{p}\right) \quad .$$

To U we associate the polynomial $H_U(x)$ with coefficients in some finite extension field L_U of K , where $H_U(x)$ is given by the expression

$$H_U(x) = \prod_{\xi \in U_x} (x - \xi) \quad .$$

We let

$$K(U)$$

be the finite extension field obtained from K by adjoining the x- and y-coordinates of all points in $U^{\#}$. We have the following obvious <u>properties</u> for H_U and $K(U)$:

(4.4) The map $U \to H_U$ is injective ($U \subset E(\mathbb{C})$, $|U| = n$).

(4.5) H_U has degree $\dfrac{n + t - 1}{2}$, where $t = |U_2|$.

(4.6) In L_U , H_U divides the n-division polynomial $T_n(E)$.

(4.7) U is invariant under the absolute Galoisgroup G over K if and only if H_U has coefficients in K .

(4.8) If U is invariant under G , then $K(U)$ is a Galois extension of K built up from the splitting field of H_U over K by adjoining some square roots. $\mathrm{Gal}(K(U),K)$ is a factor group of $G_n(E)$.

(4.9) If U is invariant under G , then H_U may uniquely be written in K as $H_U = H_1 H_2$, where H_2 is a divisor of $T_2(E)$ of degree $t = |U_2|$ and H_1 is of degree $\dfrac{n - t - 1}{2}$. We call H_2 the <u>2-part</u> of H_U .

(4.10) <u>Definition.</u> The polynomial H_U is called the <u>subgroup polynomial</u> of E corresponding to the finite subgroup U of $E(\mathbb{C})$. Any divisor of $T_n(E)$ in some finite extension field of K is called a sub-group polynomial of order n of E , if $H = H_U$ for some finite subgroup U of order n of $E(\mathbb{C})$. The set of subgroup polynomials of order n of E with coefficients in K is denoted by

$$\mathfrak{U}_n(E) \qquad .$$

In view of our above properties 4.4 and 4.7 we may identify the set of invariant subgroups of order n of $E(\mathbb{C})$ with the set $\mathfrak{U}_n(E)$. Thus the problem of finding invariant subgroups of order n of $E(\mathbb{C})$ is equivalent to the problem of finding certain divisors of $T_n(E)$ in $K[x]$ of the following degree:

$$\frac{n-1}{2} \qquad , \text{ if } n \equiv 1 \bmod 2 \ .$$

$$\frac{n}{2} \qquad , \text{ if } n \equiv 0 \bmod 2 \ , \ \frac{n}{2} \not\equiv 0 \bmod 2 \ .$$

$$\frac{n}{2} \text{ or } \frac{n}{2} + 1 \qquad , \text{ if } n \equiv 0 \bmod 4 \ .$$

(4.11) Case n is a prime. Suppose that $n = p > 2$ is a prime. Let $H \in K[x]$ be a divisor of $T_p(E)$, normed of degree $\dfrac{p-1}{2}$. Fix any root ξ of H . Put

$$p_\nu = f_\nu^2 \ , \quad q_\nu = f_{\nu-1} \, f_{\nu+1} \qquad ,$$

$\nu \in \mathbb{N}$, $\nu \geq 2$, where the f_ν's are the functions occuring in 4.2. Note that $p_\nu, q_\nu \in K[x]$. From the above discussion together with equation 4.3

we conclude that the following conditions (i) and (ii) are equivalent:

(i) $H \in \mathfrak{U}_p(E)$.

(ii) Put $\xi_1 = \xi$ and for $2 \le \nu \le \frac{p-1}{2}$ put $\xi_\nu = \frac{\xi p_\nu(\xi) - q_\nu(\xi)}{p_\nu(\xi)}$.

Then the numbers ξ_ν , $2 \le \nu \le \frac{p-1}{2}$ are well defined and ξ_ν , $1 \le \nu \le \frac{p-1}{2}$ are precisely the roots of H .

We point out that in order to check condition (ii), it is not necessary to know ξ explicitly but merely its irreducible polynomial over K . Indeed, let $G = X^d + \sum_{i=1}^{d} a_i X^{d-i}$ be the irreducible factor of H over K such that $G(\xi) = 0$. Let $H = X^q + \sum_{i=1}^{q} h_i X^{q-i}$, where $q = \frac{p-1}{2}$. For each $\nu \in \mathbb{N}$, $\nu \ge 2$ there exists a collection

$$(P_0^{(\nu)}, P_1^{(\nu)}, \ldots, P_{d-1}^{(\nu)})$$

of universal rational functions $P_\mu^{(\nu)} \in K(X_1, \ldots, X_q, Y_1, \ldots, Y_d)$ with $0 \le \mu \le d-1$ such that the above condition (ii) becomes equivalent to the following condition:

(ii') $P_\mu^{(\nu)}$ is defined at $(h_1, \ldots, h_q, a_1, \ldots, a_d)$ and $P_\mu^{(\nu)}(h_1, \ldots, h_q, a_1, \ldots, a_d) = 0$ for all $0 \le \mu \le d-1$, $2 \le \nu \le q$.

The functions $P_\mu^{(\nu)}$ may explicitly be constructed in the following way. Write $\xi_\nu = g_\nu(\xi) \in K(\xi)$ in the form $g_\nu(\xi) = \sum_{\mu=0}^{d-1} b_\mu^{(\nu)} \xi^\mu$. The coefficients $b_\mu^{(\nu)}$ are rational functions in $a_1, \ldots, a_d$ over K . Putting $H(g_\nu(\xi)) = \sum_{\mu=0}^{d-1} c_\mu^{(\nu)} \xi^\mu$ one obtains rational functions $c_\mu^{(\nu)} = P_\mu^{(\nu)}(h_1, \ldots, h_q, a_1, \ldots, a_d) \in K(h_1, \ldots, h_q, a_1, \ldots, a_d)$ for $0 \le \mu \le d-1$, and these functions clearly have the desired property.

If $H \in \mathfrak{U}_p(E)$, we conclude that $K(\xi)$ is the splitting field of H over K , of degree d , and, since obviously all irreducible factors of H over K have the same degree, d is a divisor of $\frac{p-1}{2}$. Thus, if $H = H_U$, then $K(U) = K(\xi,\chi)$, where $\chi^2 = \xi^3 + a\xi + b$, and the degree of the Galois extension $K(U)$ over K is a divisor of $p-1$.

We conclude our discussion on the group structure on an elliptic curve and come now to the main subject of this chapter. It relates finite subgroups on elliptic curves to homomorphisms of elliptic curves.

Let E and E' be elliptic curves over K . We consider homomorphisms $\lambda : E \to E'$ defined over a field L with $K \subseteq L \subseteq \mathbf{C}$. By definition such λ is a homomorphism of curves, carrying the distinguished K-rational point P on E to P' on E' . Let

$$\mathrm{Hom}_L(E,E')$$

be the set of all such homomorphisms. If E and E' are given by Weierstrass equations $y^2 = x^3 + ax + b$ resp. $Y^2 = X^3 + AX + B$ over K , then it is easy to see that the elements $\lambda \in \mathrm{Hom}_L(E,E')$ are exactly given by rational expressions of the form

$$\lambda(x,y) = \left(\frac{R(x,y)}{T(x,y)^2}, \frac{S(x,y)}{T(x,y)^3}\right) = (X,Y)$$

with polynomials $R,S,T \in L[x,y]$. If $\lambda \in \mathrm{Hom}_L(E,E')$ and if L' is any extension field of L , then λ induces an algebraic group homomorphism

$$\lambda_{L'} : E(L') \to E'(L')$$

defined over L . We consider the case $L' = \mathbb{C}$ more closely. Clearly,
$\ker(\lambda_\mathbb{C})$ consists precisely of those $P \in E(\mathbb{C})$ at which λ , given as above,
is not defined. Let $E(\mathbb{C})$ and $E'(\mathbb{C})$ analytically be represented as
$\mathbb{C}/\Lambda$ and $\mathbb{C}/\Lambda'$ resp. and let $\phi_\Lambda = (\wp_\Lambda, \wp'_\Lambda, 1)$ and $\phi_{\Lambda'} = (\wp_{\Lambda'}, \wp'_{\Lambda'}, 1)$ resp.
be the isomorphisms arising from the Weierstrass $\wp$ and $\wp'$ functions.
Let $\pi_\Lambda : \mathbb{C} \to \mathbb{C}/\Lambda$ and $\pi_{\Lambda'} : \mathbb{C} \to \mathbb{C}/\Lambda'$ resp. be the canonical homomorphisms.
Then the situation is as follows: To each $\lambda \in \mathrm{Hom}_L(E, E')$ there exists a
complex number α satisfying $\alpha\Lambda \subseteq \Lambda'$ such that the lifting $\mu_\alpha : \mathbb{C}/\Lambda \to$
$\mathbb{C}/\Lambda'$ makes the following diagram commutative:

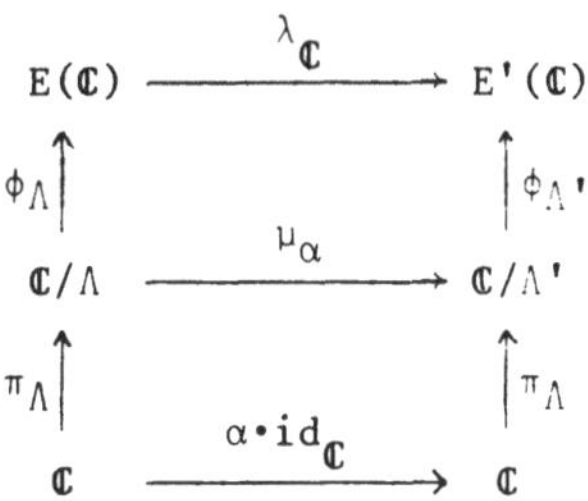

Conversely, each such $\alpha \in \mathbb{C}$ induces a homomorphism $\lambda \in \mathrm{Hom}_\mathbb{C}(E, E')$ in
this way. Using this representation of $\lambda_\mathbb{C}$ as μ_α , $\alpha \in \mathbb{C}$, it is now an
easy exercise to verify for non-zero $\lambda \in \mathrm{Hom}_L(E, E')$ the following
<u>properties</u>:

(4.12) $\ker(\lambda_\mathbb{C})$ is a finite subgroup of $E(\mathbb{C})$ and hence contained
in $E(\overline{K})$. In particular $\ker(\lambda_\mathbb{C}) = \ker(\lambda_{\overline{K}})$. ($|\ker(\lambda_\mathbb{C})|$ is
called the <u>degree</u> of λ .)

(4.13) $\lambda_\mathbb{C}$ is surjective.

(4.14) There exists $\lambda' \in \mathrm{Hom}_\mathbb{C}(E', E)$ such that $\lambda'_\mathbb{C} \cdot \lambda_\mathbb{C} = n \cdot \mathrm{id}_{E(\mathbb{C})}$
and $\lambda_\mathbb{C} \circ \lambda'_\mathbb{C} = n \cdot \mathrm{id}_{E'(\mathbb{C})}$, where $n = \deg(\lambda) = \deg(\lambda')$.
(λ' is called the <u>dual</u> of λ .) In case $L = \overline{K}$ we have in
fact $\lambda' \in \mathrm{Hom}_L(E', E)$.

The absolute Galois group G over K acts on $\mathrm{Hom}_{\overline{K}}(E,E')$ in the obvious way: For $\lambda \in \mathrm{Hom}_{\overline{K}}(E,E')$ and $\sigma \in G$ we obtain λ^σ by applying σ to the coefficients of the polynomials $R,S,T \in \overline{K}[x,y]$ defining λ . For $P \in E(\overline{K})$ we obtain

$$(\lambda^\sigma)_{\overline{K}}(P^\sigma) = \lambda_{\overline{K}}(P)^\sigma \quad .$$

Of particular interest is now

$$\mathrm{Hom}_K(E,E') \quad ,$$

the set of fixed points of G in $\mathrm{Hom}_{\overline{K}}(E,E')$. For $\lambda \in \mathrm{Hom}_K E,E')$ we have the following <u>properties:</u>

(4.15) $\ker(\lambda_{\mathbb{C}})$ (which equals $\ker(\lambda_{\overline{K}}) \subseteq E(\overline{K})$) is invariant under G .

(4.16) The dual λ' of λ is in $\mathrm{Hom}_K(E,E')$.

The ring $\mathrm{End}_K(E) = \mathrm{Hom}_K(E,E)$ obviously contains a subring isomorphic to $\mathbb{Z}$ because it contains each $\lambda^{(n)} = (g_n,h_n)$, $n \in \mathbb{Z}$, (see 4.3) where $\ker(\lambda_{\mathbb{C}}^{(n)}) = D_n(E)$. If this containment is proper, then $\mathrm{End}_K(E)$ has rank 2 over $\mathbb{Z}$.

(4.17) <u>Definition.</u> E' is called <u>isogenous</u> (over K) to E if $\mathrm{Hom}_K(E,E') \neq 0$. Any non-zero $\lambda \in \mathrm{Hom}_K(E,E')$ is called an <u>isogeny</u> (over K) from E to E' . E' is called <u>n-isogenous</u> to E , $n \in \mathbf{N}$, if there exists $\lambda \in \mathrm{Hom}_K(E,E')$ with $n = \deg(\lambda)$. We denote by

$$I_n(E)$$

the set of elliptic curves over K being n-isogenous to E and put

$$I_n^{cyc}(E) = \{E' \in I_n(E) \mid \exists \, \lambda \in Hom_K(E,E') : ker(\lambda_{\mathbb{C}}) \text{ cyclic}\} \qquad .$$

If $End_K(E)$ is isomorphic to $\mathbb{Z}$, then E is said to have <u>no complex multiplication over K</u> , if $End_L(K)$ is isomorphic to $\mathbb{Z}$ for any finite extension field L of K , then E is said to have <u>no complex multiplication.</u>

It follows now from the above discussion that the relation to be isogenous is an equivalence relation for the set of (classes of isomorphic) elliptic curves over K . Each such class of isogenous curves is simply called an <u>isogeny class.</u>

(4.18) <u>Remark.</u> Let E be an elliptic curve over K . If $n \in \mathbb{N}$ is a square, then $I_n(E)$ is clearly non-empty, since $E \in I_n(E)$. On the other hand, if $E' \in I_n(E)$, $n \in \mathbb{N}$, then $E' \in I_m^{cyc}(E)$ for some suitable $m \in \mathbb{N}$. It is now an important question which $n \in \mathbb{N}$ do imply $I_n^{cyc}(E) \neq \emptyset$. In case

$$K = \mathbb{Q}$$

examples for E with the following values of n such that $I_n^{cyc}(E) \neq \emptyset$ are known:

$$n \leq 19 \quad \text{or} \quad n \in \{21,25,27,37,43,67,163\} \qquad .$$

Mazur proved [Maz 2] that if n is a prime with $I_n^{cyc}(E) \neq \emptyset$ then n does indeed occur in the list above. In the meantime, Kenku proved [Ke 1], [Ke 2], [Ke 3], [Ke 4] that this is in fact true for arbitrary n . Moreover, Mazur proves that if $I_n(E) \neq \emptyset$, $n \in \mathbb{N}$, such that there exists E' $\in I_n(E)$ and $\lambda \in \mathrm{Hom}_K(E,E')$ with $\mathbb{Q}(\ker(\lambda_{\mathbb{C}})) = \mathbb{Q}$ or, what amounts to the same thing, such that $\ker(\lambda_{\mathbb{C}}) \subseteq E(\mathbb{Q})_{tors}$, then $n \leq 16$. More precisely, this last result can be stated as follows. If E is an elliptic curve over $\mathbb{Q}$, then the subgroup $E(\mathbb{Q})_{tors}$ is isomorphic to one of the following 15 groups:

$$\mathbb{Z}/m\mathbb{Z} \qquad \text{with} \quad 1 \leq m \leq 10 \quad \text{or} \quad m = 12 \quad ,$$
$$\mathbb{Z}/2\mathbb{Z} \times \mathbb{Z}/2n\mathbb{Z} \qquad \text{with} \quad 1 \leq n \leq 4 \quad .$$

Each of these 15 groups does indeed occur.

The proof of the following property is left as an exercise to the reader:

(4.19) Let E_1 and E_2 be elliptic curves over K with $E_1 \in I_n(E_2)$ for some $n \in \mathbb{N}$. Let E_1' and E_2' be elliptic curves over K such that E_1 and E_1' resp. E_2 and E_2' are isomorphic over a finite extension field of K . Then we have $E_1' \in I_n(E_2')$.

Next we show how for an elliptic curve E over K the sets $I_n(E)$, $n \in \mathbb{N}$, can explicitly be constructed.

(4.20) __Proposition.__ Let E be an elliptic curve over K . For each finite subgroup U of $E(\mathbb{C})$ there exists an elliptic curve E' over $K(U)$ and $\lambda \in \text{Hom}_{K(U)}(E,E')$ such that $\lambda_{\mathbb{C}}$ has kernel precisely U . The pair (E',λ) satisfies the universal mapping property for homomorphisms $\lambda \in \text{Hom}_{K(U)}(E,E'')$ with elliptic curves E'' over $K(U)$ and with $U \subseteq \ker(\lambda_{\mathbb{C}})$.

<u>Proof</u>. Fix a Weierstrass equation for E over K of the form

$$y^2 = x^3 + ax + b$$

with $a,b \in \mathcal{O}_K$. Let U in $E(\mathbb{C})$ be a finite subgroup. Let $U_2 = \{(\zeta_i,0) \mid 1 \leq i \leq t\}$ be the set of elements of order 2 in U , so that $t = t(U) \in \{0,1,3\}$. Let $s = \dfrac{n-t-1}{2}$. Then U can be written as the following disjoint union:

$$U = \{(\zeta_i,0) \mid 1 \leq i \leq t\} \cup \{(\xi_i,\pm\eta_i) \mid 1 \leq i \leq s\} \cup \{\underline{0}\}$$

with pairwise different ξ_i and with $\eta_i \neq 0$. Let E' be the elliptic curve with Weierstrass equation

$$Y^2 = X^3 + AX + B \quad ,$$

where A,B are given as follows:

$$\begin{aligned}
A &= (1-5t-10s)a - 15 \sum_{i=1}^{t} \zeta_i^2 - 30 \sum_{i=1}^{s} \xi_i^2 \\
B &= (1+21t-28s)b + 14a \sum_{i=1}^{t} \zeta_i - 14 \sum_{i=1}^{s} (5\xi_i^3+3a\xi_i)
\end{aligned} \quad . \tag{4.21}$$

We clearly have $A,B \in K(U)$. Moreover, put $\lambda(x,y) = (X,Y)$ with

$$(4.22) \qquad X = x + \sum_{i=1}^{t} \frac{3\zeta_i^2 + a}{x - \zeta_i} + \sum_{i=1}^{s} \frac{(6\xi_i^2 + 2a)(x - \xi_i) + 4\eta_i^2}{(x - \xi_i)^2}$$

$$Y = y - \sum_{i=1}^{t} \frac{(3\zeta_i^2 + a)y}{(x - \zeta_i)^2} + \sum_{i=1}^{s} \frac{(6\xi_i^2 + 2a)(x - \xi_i)y + 8\eta_i^2}{(x - \xi_i)^3} \qquad ,$$

where $\eta_i^2 = \xi_i^3 + a\xi_i + b$, $1 \leq i \leq s$. Then one can check that λ is an isogeny from E to E' . Moreover, λ is clearly defined over K(U) and
$\ker(\lambda_{\mathbb{C}}) = U$. That the pair (E',λ) satisfies the universal mapping property can easily be seen from a representation of E and E' as $\mathbb{C}/\Lambda$ resp.
$\mathbb{C}/\Lambda'$ for some suitable lattices Λ resp. Λ' in $\mathbb{C}$ and a representation
of $\lambda_{\mathbb{C}}$ as μ_α for some suitable $\alpha \in \mathbb{C}$ (see the diagram preceding 4.7).
This proves the Proposition. ∎

We write $\underline{E' = E/U}$ for the curve E' constructed in the proof of
Proposition 4.20 and call λ the $\underline{\text{canonical isogeny}}$ from E to E/U . Note
that for $U = D_n(E)$, $n \in \mathbb{N}$, we have $E/U \simeq E$ and the canonical isogeny
from E to E/U is precisely $\lambda^{(n)} = (g_n, h_n)$. There are many cases in
the classical literature where the formulas 4.21 and 4.22 appear. Mostly
they are given by expressing the Weierstrass $\wp$ and $\wp'$ functions esta-
blishing the isomorphism ϕ_Λ in terms of the $\wp$ and $\wp'$ functions esta-
blishing the isomorphism $\phi_{\Lambda'}$. See for example [Whitaker,Watson, A Course
of Modern Analysis. Cambridge University Press 1973, Chapter XX]. The formulas 4.21 and 4.22 may also be found in [Ve 1] .

Looking at the formulas 4.21 and 4.22 more closely, one sees that the
coefficients of the defining equation for E' and R,S,T , where λ is
written as $\lambda = (\frac{R(x)}{T(x)^2}, \frac{S(x,y)}{T(x)^3})$, are symmetric functions in the ζ_i ,

$1 \leq i \leq t$, and ξ_i , $1 \leq i \leq t$, resp. . From the main theorem on symmetric functions we conclude that E' and λ actually depend on the coefficients of the subgroup polynomial H_U corresponding to the finite subgroup U in $E(\mathbb{C})$. Putting all these things together we may derive the following Corollary.

(4.23) <u>Corollary.</u> Let E be an elliptic curve over K . For each $n \in \mathbb{N}$ there exists a surjective map

$$\Psi_n : \; \mathfrak{U}_n(E) \; \to \; \mathfrak{I}_n(E) \qquad . \qquad \blacksquare$$

We describe Ψ_n explicitly for $n = 2,3,5,7$. For this we fix the Weierstrass equation

$$y^2 = x^3 + ax + b$$

for E with $a,b \in \mathcal{O}_K$. First of all, we must know the coefficients of the n-division polynomials $T_n = T_n(E)$ for $n = 2,3,5,7$. They are listed in the following table.

Coefficients of the n-division polynomials T_n for $n = 2,3,5$ or 7

	T_2	T_3	T_5	T_7
x^0	b	$-a^2$	$a^6-32a^3b^2-256b^4$	$-a^{12}+160a^9b^2+3328a^6b^4+24576a^3b^6+65536b^8$
x^1	a	$12b$	$-100a^4b-640ab^3$	$392a^{10}b+7168a^7b^3+64512a^4b^5+229376ab^7$
x^2	0	$6a$	$-50a^5-240a^2b^2$	$196a^{11}+3696a^8b^2+96768a^5b^4+434176a^2b^6$
x^3	1	0	$-1600b^3-80a^3b$	$1680a^9b+152320a^6b^3+831488a^3b^5-802816b^7$
x^4		3	$-125a^4-1920ab^2$	$1302a^{10}+134400a^7b^2+394240a^4b^4-3039232ab^6$
x^5			$-696a^2b$	$57288a^8b+384517a^5b^3-3698688a^2b^5$
x^6			$-300a^3-240b^2$	$14756a^9-190400a^6b^2-2293760a^3b^4-2809856b^6$
x^7			$240ab$	$-84256a^7b-2220032a^4b^3-3809280ab^5$
x^8			$-105a^2$	$-15673a^8-831936a^5b^2-7069440a^2b^4$
x^9			$380b$	$-425712a^6b-3727360a^3b^3-1555456b^5$
x^{10}			$62a$	$-42168a^7-1192800a^4b^2-3362816ab^4$
x^{11}			0	$-608160a^5b-2603776a^2b^3$
x^{12}			5	$-111916a^6-297472a^3b^2-928256b^4$
x^{13}				$-161840a^4b-2132480ab^3$
x^{14}				$-82264a^5-615360a^2b^2$
x^{15}				$-31808a^3b-829696b^3$
x^{16}				$-35231a^4-571872ab^2$
x^{17}				$92568a^2b$
x^{18}				$-19852a^3-42896b^2$
x^{19}				$-112ab$
x^{20}				$-2954a^2$
x^{21}				$3944b$
x^{22}				$308a$
x^{23}				0
x^{24}				7

Further we must know the functions $g_\nu = g_\nu(E)$ for $\nu = 2,3$:

$$g_2(x,y) = \frac{x^4 - 2ax^2 - 8bx + a^2}{4(x^3 + ax + b)}$$

$$g_3(x,y) =$$

$$\frac{x^9 - 12ax^7 - 96bx^6 + 30a^2x^5 - 24abx^4 + (36a^3 + 48b^2)x^3 + 48a^2bx^2 + (9a^4 + 96ab^2)x + 64b^3 + 8a^3b}{9x^8 + 36ax^6 + 72bx^5 + 30a^2x^4 + 144abx^3 + (144b^2 - 12a^3)x^2 - 24a^2bx + a^4}$$

(4.24) The case $n = 2$. The subgroup polynomials in $\mathrm{U}_2(E)$ are exactly the normed linear factors $H = x - \zeta$ in $K[x]$ of $T_2(E) = x^3 + ax + b$. For such H we obtain $E' = \Psi_2(H)$ as follows. E' has a Weierstrass $Y^2 = X^3 + AX + B$ with

$$\boxed{\begin{aligned} A &= -4a + 15\zeta^2 \\ B &= 22b + 14a\zeta \end{aligned}} \quad .$$

The canonical isogeny λ of degree 2 from E to E' is given by $\lambda(x,y) = (X,Y)$ with

$$X = x + \frac{3\zeta^2 + a}{x - \zeta} \quad , \quad Y = y - \frac{(3\zeta^2 + a)y}{(x - \zeta)^2} \quad .$$

Suppose $b = 0$. Then we always have $H = x \in \mathrm{U}_2(E)$. Hence the elliptic curve E' over K with Weierstrass equation

$$Y^2 = X^3 - 4aX$$

is 2-isogenous to E with an isogeny of degree 2 as follows:

$$X = x + \frac{a}{x} \quad , \quad Y = y - \frac{ay}{x^2} \quad .$$

The other two subgroup polynomials of order 2 in this case are $H = x \pm \sqrt{-a}$.

(4.25) The case $n = 3$. The subgroup polynomials in $\mathfrak{U}_3(E)$ are exactly the normed linear factors $H = x - \xi$ in $K[x]$ of $T_3(E) = 3x^4 + 6ax^2 + 12bx - a^2$. For such H we obtain $E' = \Psi_3(H)$ as follows. E' has a Weierstrass equation $Y^2 = X^3 + AX + B$ with

$$\boxed{\begin{aligned} A &= -3(3a + 10\xi^2) \\ B &= -(70\xi^3 + 42a\xi + 27b) \end{aligned}} \qquad .$$

The canonical isogeny λ of degree 3 from E to E' is given by $\lambda(x,y) = (X,Y)$ with

$$X = x + \frac{(6\xi^2+2a)(x-\xi) + 4\chi^2}{(x-\xi)^2} \quad , \quad Y = y - \frac{(6\xi^2+2a)(x-\xi)y + 8\chi^2}{(x-\xi)^3} \quad ,$$

where $\chi^2 = \xi^3 + a\xi + b$. Suppose $a = 0$. Then we always have $H = x \in \mathfrak{U}_3(E)$. Hence the elliptic curve E' over K with Weierstrass equation

$$Y^2 = X^3 - 3^3 b$$

is 3-isogenous to E with an isogeny of degree 3 as follows:

$$X = x + \frac{4b}{x^2} \quad , \quad Y = y - \frac{8by}{x^3} \quad .$$

The other three subgroup polynomials of order 3 of E are in this case given by $H = x - \rho^i \alpha$, $0 \le i \le 2$, where α is any cube root of $-4b$ and ρ is a primitive cube root of unity.

(4.26) **The case $n = 5$** . The subgroup polynomials in $\amalg_5(E)$ are exactly the quadratic divisors $H = x^2 + h_1 x + h_2$ in $K[x]$ of $T_5(E)$ satisfying the following condition. If $\xi = \xi_1$ is one of the roots of H then

$$\xi_2 = g_2(\xi) = \frac{\xi^4 - 2a\xi^2 - 8b\xi + a^2}{4(\xi^3 + a\xi + b)}$$

is the other root of H . For such H we obtain $E' = T_5(H)$ as follows. E' has a Weierstrass equation $Y^2 = X^3 + AX + B$ with

$$
\begin{aligned}
A &= -19a - 30(h_1^2 - 2h_2) \\
B &= -55b - 14(15h_1h_2 - 5h_1^3 - 3ah_1)
\end{aligned}
$$

Example. Put $a = 6$. $b = -88$. Then E is the curve 144*i(1+i) over $\mathbb{Q}(i)$ in the appendix of these notes. The coefficients of $T_5(E)$ are as follows:

Coefficient of x^i

i :				
0	-15 405 681 088		6	-1 923 360
1	2 628 257 280		7	-126 720
2	-67 296 960		8	-3 780
3	1 091 875 840		9	-33 440
4	-89 372 880		10	372
5	2 204 928		12	5

Checking the quadratic divisors of $T_5(E)$ by means of a computer, one obtains that $\amalg_5(E)$ contains exactly the following element:

$$H = x^2 + 4x - 14 \quad .$$

The elliptic curve $E' = \Psi_5(H)$ has the Weierstrass equation

$$Y^2 = X^3 - 1434X + 22088 \qquad .$$

E' is the curve $160*i(1+i)$ over $\mathbb{Q}(i)$ in the appendix.

(4.28) The case $n = 7$. The subgroup polynomials in $\mathbb{U}_7(E)$ are exactly the cubic divisors $H = x^3 + h_1 x^2 + h_2 x + h_3$ in $K[x]$ of $T_7(E)$ satisfying the following condition. Fix any root $\xi = \xi_1$ of H . Then

$$\xi_2 = g_2(\xi) \quad , \quad \xi_3 = g_3(\xi)$$

are the other roots of H . For such H we obtain $E' = \Psi_7(H)$ as follows. E' has a Weierstrass equation $Y^2 = X^3 + AX + B$ with

$$\boxed{\begin{aligned} A &= -29a - 30(h_1^2 - 2h_2) \\ B &= -83b - 14(3h_1 h_2 - h_1^3 - 3h_3) \end{aligned}} \qquad .$$

We conclude now the discussion on the explicit construction of isogenous curves for the moment and consider numerical functions on the set of elliptic curves over K which are constant on isogeny classes.

Let E and E' be elliptic curves over K such that E and E' are isogenous (over K). Then we have the following <u>properties</u>:

(4.28) E and E' have the same conductor.

(4.29) Let $\mathfrak{p}$ be any prime ideal in O_K . Then for the reduction of
E resp. E' at $\mathfrak{p}$ the following equality holds:

$$|E_{\mathfrak{p}}(k_{\mathfrak{p}})| = |E'_{\mathfrak{p}}(k_{\mathfrak{p}})| \quad .$$

For a proof of property 4.28 we refer to [Neu,I;p.110] . A proof of property
4.29 may be found, for example, in [Ca;p.242] together with [Lan 1;p.112] ,
in case $\mathfrak{p}$ does not divide Cond(E) . If $\mathfrak{p}$ does divide the conductor
of E then property 4.29 follows from property 4.28 (see chapter 1, the
discussion preceding Definition 1.12). Note that if

$$y^2 = x^3 + ax + b$$

with $a,b \in O_K$ is a Weierstrass equation for E over K , then $|E_{\mathfrak{p}}(k_{\mathfrak{p}})|$
is given by

$$|E_{\mathfrak{p}}(k_{\mathfrak{p}})| = 1 + \alpha_{\mathfrak{p}} + 2\beta_{\mathfrak{p}} \quad ,$$

where $\alpha_{\mathfrak{p}}$ denotes the number of distinct roots and $\beta_{\mathfrak{p}}$ the number of distinct non-zero squares of $x^3 + ax + b$ in O_K modulo $\mathfrak{p}$. We also point
out that if E and E' are isogenous over K , then by property 4.29
the L-series of E and E' (see 1.13) coincide. It has been conjectured
that the converse of this statement is true (isogeny conjecture).

Now let $\mathfrak{a}$ be a given ideal in O_K . The determination of all elliptic curves over K with conductor $\mathfrak{a}$ can algorithmically be carried out.
This determination corresponds to the first two of our three steps in the
introduction, where S is now the set of prime ideals in O_K dividing $\mathfrak{a}$.

In chapter 2 and 3 we showed how to carry these steps explicitly out in case O_K is a principal ideal domain. We consider now the last of our three steps:

Find for the set of elliptic curves over K with conductor $\mathfrak{a}$
the division into isogeny classes.

We first want to discuss to what extent this last step can algorithmically be carried out. For this we have to consider the following related problems:

(A) Given an elliptic curve E over K , find all elliptic curves over K being isogenous to E .

(B) Given two elliptic curves over K , decide whether they are isogenous (over K) or not.

Obviously, an algorithm solving problem (A) or problem (B) yields an algorithm for carrying out our above step. Moreover, using the algorithm for determining all elliptic curves over K with given conductor, problem (A) can algorithmically be solved if and only if this is true for problem (B).

Now in case $K = \mathbb{Q}$ an algorithm solving problem (A) is given as follows. For the determination of all elliptic curves over K , being isogenous to E , it is sufficient to determine $\mathcal{I}_n^{cyc}(E)$, $n \in \mathbb{N}$. From Mazur's result (resp. its generalization to arbitrary n ; see remark 4.18) we know that $\mathcal{I}_n^{cyc}(E) \neq \emptyset$ is valid only for

$$n \leq 19 \quad \text{or} \quad n \in \{21,25,27,37,43,67,163\} \quad .$$

On the other hand, for each such n , $\mathcal{I}_n^{cyc}(E)$ can effectively be determined by Corollary 4.23. This gives the desired algorithm in case $K = \mathbb{Q}$ and hence a solution of problem (A). Of course, algorithmical solutions

of problems (A) and (B) should also exist for arbitrary K . However for
any $K \neq \mathbb{Q}$ such solutions are unknown.

We now discuss the <u>practical implementation</u> of our above step. The
algorithm described above for carrying out our step in case $K = \mathbb{Q}$ cannot
be used for explicit calculations since the map Ψ_n in Corollary 4.23 can
explicitly be given only for small values of n . So in practice one pro-
ceeds for $K = \mathbb{Q}$ and also for arbitrary K as follows.

(i) For each curve E in the list of elliptic curves over K with
the given conductor calculate

$$|E_{\mathfrak{p}}(k_{\mathfrak{p}})|$$

for a few prime ideals $\mathfrak{p}$ with small norms.

(ii) Arrange the curves in the list into classes such that two curves
of the list are in the same class if and only if the above number is the
same for each of the prime ideals $\mathfrak{p}$. In view of property 4.29 each such
class C is a union of isogeny classes. Now C should in fact be a full
isogeny class. In many instances this turns really out to be true and one
can check this in the following way.

(iii) First of all it is convenient to introduce the <u>isogeny graph</u>
<u>of weight n associated to</u> C . Let $n \in \mathbb{N}$ be squarefree. The elements of
C become the vertices of the graph. Different vertices E and E' of the
graph are connected by an edge of weight p , p a rational prime,

- 80 -

$$E \bullet \!\!\xrightarrow{\ \ p\ \ }\!\! \bullet E'$$

if and only if $E' \in I_p(E)$ (and hence $E \in I_p(E')$) and p divides n . Now it is obvious that C is a full isogeny class if and only if for some squarefree $n \in N$ the isogeny graph of weight n associated to C is connected. The construction of the isogeny graph of weight n amounts to the determination of $I_p(E)$ for each $E \in C$ and each prime p dividing n . For this we refer to Corollary 4.23 and the subsequent calculations for $p \in \{2,3,5,7\}$. In many instances of C it turns out that the isogeny graph of weight $2^{\alpha_1}3^{\alpha_2}5^{\alpha_3}7^{\alpha_4}$ for some $\alpha_i \in \{0,1\}$, $1 \le i \le 4$ is connected and one is finished. For example, if this is the case for $|C| \le 4$, then the isogeny graph of weight 6 associated to C is one of the following 7 types:

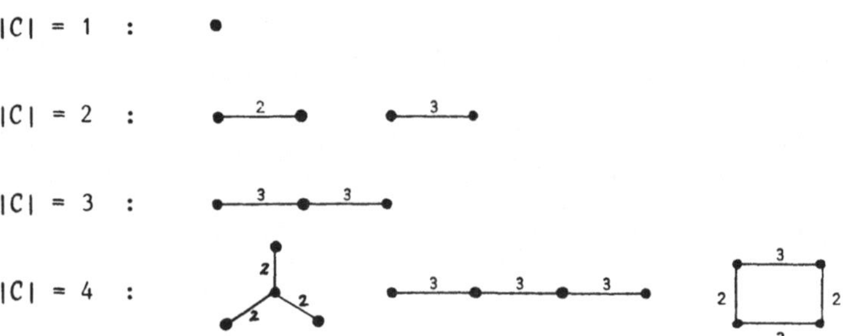

If it is necessary to construct the isogeny graph of weight n , for some n with a prime divisor greater than 7 , in order to obtain a connected graph, then the present method for explicitly carrying out our last step seems unfortunately to be impracticable.

We conclude this chapter and state without proof some _properties_ of the isogeny graph of weight n , $n \in N$, of an isogeny class C , where we assume that C contains an elliptic curve without complex multiplication over K .

(4.30)   All isogenies between elliptic curves in C have the same degree.

(4.31)   For every $n \in \mathbb{N}$ the isogeny graph of weight n is a finite tree.

(4.32)   There is an isogeny graph of weight n with minimal n such that every isogeny graph of weight m is a subgraph.

The verification of these properties is left as an exercise to the reader.

Let $K = \mathbb{Q}$. From Mazur's and Kenku's classification of the set of numbers $n \in \mathbb{N}$ for which $I_n^{cyc}(E) \neq \emptyset$ is possible (see remark 2.8) one may derive the following result: If C is an arbitrary isogeny class then

(4.33) $|C| \leq 8$.

(See [Kenku, J. Number Theory 15, 199-202 (1982)].)

Chapter 5

Review on Explicit Results

The set $E(S)$ of all elliptic curves over K with good reduction outside a finite set of prime ideals S in 0_K and the division into isogeny
classes has explicitly been determined for several concrete choices of K
and S . In most cases $E(S)$ is obtained by implementing the three steps
stated in the introduction. In this chapter we give a review on these and
related results.

(5.1)   The case $K = Q$ and $|S| = 1$. Suppose that $S = \{2\}$. We saw
in 2.11 (see also example (1) of 2.3) that in order to construct $E(\{2\})$
one has to determine all basic solutions (p,q,ε,m,n) with $q \geqq 0$ of the
diophantine equation

$$x^3 - y^2 = \xi \, 2^{z_1} 3^{z_2} \quad ,$$

$x,y \in \mathbb{Z}$, $\xi \in \{1,-1\}$, $z_1,z_2 \in \mathbb{N}_0$, with $n = 3$ such that a $\Gamma_{v,w}$ exists
for some $\mu,\alpha \in \{0,1\}$, $\gamma \in \{0,1,2\}$, where

$$v = (-1)^{2\mu} 2^{4+2\alpha-4\gamma} p$$
$$w = (-1)^{3\mu} 2^{6+3\alpha-6\gamma} q \quad .$$

For each such basic solution one obtains an elliptic curve in $E(\{2\})$ with
Weierstrass equation

$$y^2 = x^3 - 3\lambda^2 px - 2\lambda^3 q \qquad ,$$

where $\lambda = (-1)^\mu 2^\alpha 3$ and conversely, each elliptic curve in $E(\{2\})$ arises
uniquely in this way. A necessary condition for such a basic solution is
the following: 3 divides q implies 3^3 divides q . Actually Ogg [Ogg 2]
constructed $E(\{2\})$ in a slightly different way. Indeed, if S is any
given finite set of primes in $\mathbb{Z}$, the construction of the subset of all
ellitpic curves in $E(S)$ having a non-trivial torsion subgroup is connected
with a certain diophantine problem similar to that of chapter 2 . The present problem however is of a much simpler nature. It can be solved in a
fairly elementary way. Taking this into account, Ogg proceeded in the
following two steps. In the first step he showed that there is no elliptic
curve over $\mathbb{Q}$ with good reduction everywhere (i.e. $E(\emptyset) = \emptyset$) and each
elliptic curve in $E(\{2\})$ has additive reduction at 2 . In the second
step he showed that each elliptic curve in $E(\{2\})$ has a $\mathbb{Q}$-rational
point of order 2 . The solution of the associated diophantine problem gives
$E(\{2\})$. There are precisely 24 curves in $E(\{2\})$. (See also [Neu 1, p.308].)

The solution of a similar diophantine problem enables Miyawaki [Mi]
to prove the following result: There are precisely 16 elliptic curves
over $\mathbb{Q}$ such that $E \in E(S)$ for some S with $|S| = 1$, $S \neq \{2\}$ and
such that $|E(\mathbb{Q})_{tors}| \geqq 3$. These curves are listed at the end of his paper.

Miyawaki's result is extended in papers of Hadano and Ogg. From these
papers the (finite) set of elliptic curves E over $\mathbb{Q}$ is known with
$E \in E(S)$ such that one of the following three conditions holds:

(i) $|S| = 2$, $2 \in S$, $|E(\mathbf{Q})_{tors}| = 2,3$ or 5 . (See [Ha 1],[Ha 2]

   and [Ogg 3].)

(ii) $|S| = 2$, $|E(\mathbf{Q})_{tors}| = 5$. (See [Ha 2].)

(iii) $|S| = 3$, $2 \in S$, $|E(\mathbf{Q})_{tors}| \geq 6$. (See [Ha 2].)

(5.2) <u>The case $K = \mathbf{Q}$ and $S = \{2,3\}$</u> . Consider the diophantine
equation

$$x^3 - y^2 = \xi \, 2^{z_1} 3^{z_2}$$

with $x,y \in \mathbf{Z}$, $\xi \in \{1,-1\}$, $z_1,z_2 \in \mathbf{N}_0$. Each basic solution $\beta = (p,q,\varepsilon,$
$m,n)$ with $q \geq 0$ defines a series of 8 pairwise non-isomorphic elliptic
curves $\beta * \lambda$ over $\mathbf{Q}$, $\lambda = (-1)^{\mu} 2^{\alpha} 3^{\beta}$, $\mu,\alpha,\beta \in \{0,1\}$, each such $\beta * \lambda$ with
Weierstrass equation

$$y^2 = x^3 - 3\lambda^2 px - 2\lambda^3 q \quad .$$

By Theorem 2.4 (see also example (1) of 2.3) this construction precisely
yields the elliptic curves in $E(\{2,3\})$. Now the set of all such basic
solutions of the above equation was determined by Coghlan ([Cog 1], [Cog 2]).
He parametrizes the solutions of the above equation by the solutions of
certain binary cubic resp. quartic equations of Thue type over $\mathbf{Z}$ (see
Lemma 3.2 and the discussion after 3.6) and he solves that equations.
There are precisely 95 basic solutions of the above equation (with
$q \geq 0$) and hence $|E(\{2,3\})| = 760$. Among the curves

in $E(\{2,3\})$ are the curves in [Ogg 2] and also the curves in [Ogg 3] (resp. [Neu 2]) which are obtained amomg others by an analysis of the 2-division field. The curves in $E(\{2,3\})$ and the isogeny classes are given in [Modular Functions of One Variable IV. Springer LN 476(1975), 123-134].

(5.3) <u>The case $K = \mathbb{Q}(i)$ and $S = \{1+i\}$</u> . We saw in 2.11 (see also example (2) of 2.3) that in order to construct $E(\{1+i\})$ one has to determine all basic solutions (p,q,ε,m,n) with $\mathrm{Re}(q) \geq 0$ and $\varepsilon \in \{1,i\}$, where $\mathrm{Im}(q) \geq 0$ in case $\mathrm{Re}(q) = 0$, of the diophantine equation

$$x^3 - y^2 = \xi\,(1+i)^{z_1}\,3^{z_2} \qquad ,$$

$x,y \in \mathbb{Z}[i]$, $\xi \in \{1,-1,i,-i\}$, $z_1,z_2 \in \mathbb{N}_0$ with $n = 3$ such that a $\Gamma_{v,w}$ exists for some $\mu,\alpha \in \{0,1\}$, $\gamma \in \{0,1,2\}$, where

$$v = i^{2\mu}\,(1+i)^{8+2\alpha-4\gamma}\,p$$
$$w = -i^{3\mu}\,(1+i)^{12+3\alpha-6\gamma}\,q \qquad .$$

To each such basic solution an elliptic curve in $E(\{1+i\})$ is attached, with Weierstrass equation

$$y^2 = x^3 - 3\lambda^2 px - 2\lambda^3 q \qquad ,$$

where $\lambda = (-1)^\mu(1+i)^\alpha 3$ and conversely, each elliptic curve in $E(\{1+i\})$ arises uniquely in this way. Such a basic solution satisfies the following condition: 3 divides q implies 3^3 divides q . In his thesis [Stro 1], Stroeker completely determined all the above (so called "good") basic solutions by means of the methods described in chapter 3. In this way he obtains all elliptic curves in $E(\{1+i\})$. There are precisely 64 such curves. They can be found in the lists in the appendix of these notes.

(5.4)   The case $K = \mathbb{Q}(i)$ and $S = \{1+i,3\}$. We report upon this case more closely. Consider the diophantine equation

$$x^3 - y^2 = \xi \, (1+i)^{z_1} \, 3^{z_2}$$

with $x,y \in \mathbb{Z}[i]$, $\xi \in \{1,-1,i,-i\}$, $z_1,z_2 \in \mathbb{N}_0$. Each basic solution $\beta = (p,q,\varepsilon,m,n)$ with $\mathrm{Re}(q) \geqq 0$ and $\varepsilon \in \{1,i\}$, where $\mathrm{Im}(q) \geqq 0$ in case $\mathrm{Re}(q) = 0$, defines a series of 8 pairwise non-isomorphic curves $\beta * \lambda$ over $\mathbb{Q}(i)$, $\lambda = i^{\mu}(1+i)^{\alpha}3^{\beta}$, $\mu,\alpha,\beta \in \{0,1\}$, each such $\beta * \lambda$ with Weierstrass equation

$$y^2 = x^3 - 3\lambda^2 px - 2\lambda^3 q \qquad .$$

By Theorem 2.4 (see also example (2) of 2.3) this construction precisely yields all elliptic curves in $E(\{1+i,3\})$. In [Las 2] all basic solutions of the above equation are determined by means of the methods described in chapter 3, up to a certain assumption which however should be true in view of great numerical evidence. We are going to explain this briefly.

We put $O = \mathbb{Z}[i]$ and $R = O[\frac{1}{1+i},\frac{1}{3}]$. First of all, by Lemma 3.2 the set B_S of all basic solutions $\beta = (p,q,\varepsilon,m,n)$ with $\mathrm{Re}(q) \geqq 0$ and $\varepsilon \in \{1,i\}$, where $\mathrm{Im}(q) \geqq 0$ in case $\mathrm{Re}(q) = 0$, are in one-to-one corres-pondence with the union of the sets of solutions in R of the equations

$$\Gamma_r : \quad x^3 - y^2 = r \quad ,$$

where $r = -i^j(1+i)^k 3^l$, $j \in \{0,1\}$, $0 \leq k,l \leq 5$. For each such r there is a finite set $K_r \subseteq \mathbb{Q}(i) \times \mathbb{Q}(i)$ and for each $(\sigma,\tau) \in K_r$ a binary cubic form $f_{\sigma,\tau}$, given by 3.7, such that the solutions in R can be parametrized by the solutions $(X,Y) \in \Delta^{-1}R \times \Delta^{-1}R$ of the equations

$$f_{\sigma,\tau}(X,Y) = s \quad , \quad (\sigma,\tau) \in K_r \quad ,$$

with some suitable $s = s_r \in O$ and $\Delta = \Delta_r \in O$ (see Theorem 3.14 and 3.19). For each r the set K_r , s_r and Δ_r are explicitly given in example 3.20. In particular we have $\Delta_r = 2$ in case $j = 0$, k even, l odd and $\Delta_r = 1$ otherwise. In any case we have $\Delta^{-1}R = R$. We have $s_r = i^{\gamma_r}(1+i)^{a_1}3^{b_1}$ for some suitable $a_1,b_1 \in \mathbb{N}_0$, $0 \leq a_1,b_1 \leq 2$, where $\gamma_r = 1$ in case $j = 0$, k even, l odd and $\gamma_r = 0$ otherwise. Moreover, each $f_{\sigma,\tau}$, $(\sigma,\tau) \in K_r$ has coefficients in O beside the following two cases of r : (i) $j = 0$, k even, l odd and (ii) $j = 0$, k even, l even. In these cases each $f_{\sigma,\tau}$, $(\sigma,\tau) \in K_r$ has coefficients in $\frac{1}{2}O$. This alltogether leads to a list of 21 binary cubic forms $f(X,Y)$ with coefficients in $\frac{1}{2}O$ and for each form f in the list the solutions (X,Y) with $X,Y \in R$ of

$$f(X,Y) = (1+i)^{a_1}3^{b_1} \quad \text{resp.} \quad i(1+i)^{a_1}3^{b_1} \quad , \quad 0 \leq a_1,b_1 \leq 2$$

are to be determined. We put $\alpha_f = 0$ in case f has coefficients in O and $\alpha_f = 1$ otherwise. Then it is easy to see that it is sufficient to determine for each f in the list all $X,Y \in O$ such that $\gcd(X,Y)$ divides $(1+i)^{\alpha_f}$ and such that

$$f(X,Y) = (1+i)^a 3^b \quad \text{resp.} \quad i(1+i)^a 3^b$$

for some $(a,b) \in \mathbb{N}_0 \times \mathbb{N}_0$.

On the other hand, for 26 of the above 72 values of r one is able to determine $\Gamma_r(R)$ completely by an analysis of the fibres of the "trace map", introduced in chapter 3:

$$\sigma_{r,R} : \quad \Gamma_r(R) \rightarrow \Gamma_r(Q) \cup \{\underline{0}\} \quad .$$

This result may be found in [Las 3]. An important consequence of this result obviously is that the 21 equations $f(X,Y) = (1+i)^a 3^b$ resp. $i(1+i)^a 3^b$ need not all be solved completely. Indeed, for each f one can a and b assume to be of a certain restricted form and moreover, 3 of these equations have to be considered not at all. 7 of the remaining 18 equations have been solved completely. For the other 11 equations the form f and the associated $a,b \in \mathbb{N}_0$ are as follows:

1. $X^3 + 3(1+i)X^2Y + 9iXY^2 + 3i(1+i)Y^3$, $a \equiv \mu \bmod 3$, $b \equiv \nu \bmod 3$ with
$$(\mu,\nu) = (0,2),(1,1),(1,2) \text{ or } (2,2) .$$

2. $X^3 + 3X^2Y + 3(1+i)XY^2 + (1+i)Y^3$, $a,b \in \mathbb{N}_0$.

3. $X^3 + 3(2+i)X^2Y + 9(1+i)XY^2 + 3(1+3i)Y^3$, $a,b \in \mathbb{N}_0$.

4. $\dfrac{-1+i}{2}X^3 + 3\dfrac{1+i}{2}X^2Y + 9\dfrac{-1+i}{2}XY^2 + 3\dfrac{1+i}{2}Y^3$

5. $-\dfrac{1}{2}X^3 - \dfrac{3}{2}X^2Y - \dfrac{9}{2}XY^2 - \dfrac{3}{2}Y^3$

6. $-\dfrac{i}{2}X^3 + 3\dfrac{-2+i}{2}X^2Y - 9\dfrac{i}{2}XY^2 + 3\dfrac{-2+i}{2}Y^3$

7. $-\dfrac{i}{2}X^3 + 3\dfrac{2+i}{2}X^2Y - 9\dfrac{i}{2}XY^2 + 3\dfrac{2+i}{2}Y^3$

$\left.\begin{array}{l} \\ \\ \\ \\ \\ \end{array}\right\}$ $a \equiv 1 \bmod 3$, $b \equiv 0 \bmod 3$.

$$8. \quad (1-i)X^3+3X^2Y+3(1+i)XY^2+iY^3$$

$$9. \quad 3X^3-9i(1+i)X^2Y+9iXY^2+3(1+i)Y^3$$

$$10. \quad 3(1-2i)X^3+18X^2Y+9(2+i)XY^2+6iY^3$$

$$11. \quad -3(1+2i)X^3-18iX^2Y+9(2-i)XY^2+6Y^3$$

$a \equiv \mu \bmod 3$, $b \equiv \nu \bmod 3$ with

$(\mu,\nu) = (0,1),(1,0),(1,1)$ or $(2,1)$.

Now, if there should exist a yet unknown solution (X,Y) with $X,Y \in 0$ such that $\gcd(X,Y)$ divides $(1+i)^\alpha f$, then in each of the above cases (X,Y,a,b) must satisfy the following conditions (N denotes here the norm map of $\mathbb{Q}(i)$ over $\mathbb{Q}$):

Equ. 1 : $a \equiv 1 \bmod 3$, $b = 1$ and $N(X) > 10^5$ or $N(Y) > 2500$.

Equ. 2 : $N(X)$ or $N(Y) > 2500$.

Equ. 3 : $b = 0$ or 1 and $N(X)$ or $N(Y) > 2500$.

Equ. 4 : $a \equiv 1 \bmod 3$, $b = 0$ and $N(X)$ or $N(Y) > 2500$.

Equ. 5 :

Equ. 6 : $\Big\}$ $a = 1$, $b = 0$ and $N(X)$ or $N(Y) > 2500$.

Equ. 7 :

Equ. 8 : $N(X)$ or $N(Y) > 2500$.

Equ. 9 :

Equ.10 : $\Big\}$ $b = 1$ and $N(X)$ or $N(Y) > 2500$.

Equ.11 :

It seems very unlikely that such solutions do exist. For example, all norms $N(X),N(Y)$ for the solutions (X,Y) which are actually known are very small in comparison with the ranges given above.

One obtains precisely 160 basic solutions in B_S and hence 1280 curves in $E(\{1+i,3\})$. The basic solutions are listed in table 1 of the appendix. The isogeny classes for the elliptic curves in $E(\{1+i,3\})$ [Las 4] may be determined as described in chapter 4. The curves and the division into the isogeny classes are given in table 2 and 3 of the appendix.

(5.5)   The case $K = \mathbb{Q}(\sqrt{-2})$ and $S = \{\sqrt{-2}\}$. The construction of $E(\{\sqrt{-2}\})$ amounts to the determination of all basic solutions (p,q,ε,m,n) with $\mathrm{Re}(q) \geq 0$ and, if $\mathrm{Re}(q) = 0$, with $\mathrm{Im}(q) \geq 0$ of the diophantine equation

$$x^3 - y^2 = \xi(\sqrt{-2})^{z_1}3^{z_2} \quad ,$$

$x,y \in \mathbb{Z}[\sqrt{-2}]$, $\xi \in \{1,-1\}$, $z_1,z_2 \in \mathbf{N}_0$, with $n = 3$ such that a $\Gamma_{v,w}$ exists for some $\mu,\alpha \in \{0,1\}$, $\gamma \in \{0,1,2\}$, where

$$v = (-1)^{2\mu}(\sqrt{-2})^{8+2\alpha-4\gamma}p$$
$$w = (-1)^{3\mu}(\sqrt{-2})^{12+3\alpha-6\gamma}q \quad .$$

To each such basic solution an elliptic curve in $E(\{\sqrt{-2}\})$ is attached, given by the Weierstrass equation

$$y^2 = x^3 - 3\lambda^2 px - 2\lambda^3 q \quad ,$$

where $\lambda = (-1)^\mu(\sqrt{-2})^\alpha 3$ and conversely, each elliptic curve in $E(\{\sqrt{-2}\})$ arises uniquely in this way. For all this we refer to 2.11. In his thesis [Stro 1], Stroeker completely determined all the above (so called "good") basic solutions by means of the methods described in chapter 3. In this way he obtains all elliptic curves in $E(\{\sqrt{-2}\})$. There are precisely 40 such curves.

(5.6) <u>The case K quadratic and S = $\emptyset$</u> . It is an important question
which number fields K other than $\mathbb{Q}$ have the property $E(\emptyset) = \emptyset$. We put

$$E_K = E(\emptyset)$$

for the set of elliptic curves over K with good reduction everywhere. If
K has class number one, then for the study of E_K one can use the results
of chapter 2. In this case one has to consider the diophantine equation

$$x^3 - y^2 = \xi\,\pi_1^{z_1}\cdots\pi_n^{z_n} \quad ,$$

where $S = \{\pi_1,\ldots,\pi_n\}$ are exactly the prime divisors of 2 and 3 (up
to associates). Let B be a set of non-associated basic solutions of that
equation. Then each $E \in E(S)$ is of the form $E = \beta * \lambda$ for some $\beta =$
$(p,q,\epsilon,m_1,\ldots,m_n) \in B$ and $\lambda \in \mathbf{R} = \{\zeta^\mu \eta^\nu \pi_1^{\alpha_1}\cdots\pi_n^{\alpha_n} \mid \mu,\nu,\alpha_1,\ldots,\alpha_n = 0 \text{ or } 1\}$.
If now E has good reduction at all primes π_i , i.e. $E \in E_K$, then
both Corollaries 2.10 and 2.11 give restrictions for the solution β and
in fact, by means of the methods described in chapter 3 one can often show
that such solutions cannot exist. Stroeker [Stro 1] has shown that this is
indeed the case if K is imaginary quadratic (with class number one) so
that $E_K = \emptyset$ for such K . Stroeker actually proved (main theorem in
[Stro 1]) that if K is an imaginary quadratic field and $E \in E_K$, then
E has no global minimal Weierstrass equation over K .

Further results concerning E_K for imaginary quadratic fields may be
found in Setzer's paper [Set 2]. He obtains his result by analyzing the
normal closure over $\mathbb{Q}$ of the 2-division field and using class field theory.

$$- 92 -$$

These results are as follows. Let $K = \mathbb{Q}(\sqrt{-m})$, $m \in \mathbb{N}$ squarefree.

(i) There is an elliptic curve $E \in E_K$ having a point of order 2 in $E(K)$ if and only if

$$m = 65\, m_1 \qquad ,$$

where m_1 is a square $\bmod 5$ and $\bmod 13$ and 65 is a square $\bmod m_1$. If this is the case then the number of such curves in E_K is 2^r , where r is the number of primes in $\mathbb{Z}$ which ramify in the extension $K : \mathbb{Q}$.

(ii) $E_K = \emptyset$ for the following values of m : $1,2,3,5,6,7,10,13,14,15,$
$17,21,22,30,33,34,39,41,42,46,47,55,57,58,62,66,69,70,73,77,78,82,85,$
$86,93,94,95,97,102,103,105,113,114,119,122,130,133,134,137,138,$
$143,145,146,149,151,154,159,161.$

An example for an imaginary quadratic field K such that $E_K \neq \emptyset$ may be obtained from the result (i) above:

$$E_{\mathbb{Q}(\sqrt{-65})} \neq \emptyset \qquad .$$

$E_{\mathbb{Q}(\sqrt{-65})}$ contains at least 8 elliptic curves, each of them with a point of order 2 defined over $\mathbb{Q}(\sqrt{-65})$.

Examples for real quadratic fields K and global minimal Weierstrass equations for elliptic curves over K in E_K are given as follows:

$$K = \mathbb{Q}(\sqrt{7}) . \qquad y^2 + xy = x^3 - 2\varepsilon x^2 + \varepsilon^2 x \qquad , \text{ with } \quad \varepsilon = 8 + 3\sqrt{7} ,$$
$$\text{a fundamental unit in } K .$$

$K = \mathbb{Q}(\sqrt{29})$. $\qquad y^2 + xy + \varepsilon^2 y = x^3 \qquad$, with $\varepsilon = \frac{1}{2}(5 + \sqrt{29})$, a fundamental unit in K .

$K = \mathbb{Q}(\sqrt{41})$. $\qquad y^2 + xy = x^3 - \varepsilon x \qquad$, with $\varepsilon = 32 + 5\sqrt{41}$, a fundamental unit in K .

Further results for real quadratic fields may be found in Setzer's recent paper [Set 3]. For imaginary quadratic fields we also refer to [Is].

(5.7) Concluding remarks. We finally mention the papers [Set 1], [Bö] and [Bru&Kra] in which elliptic curves over $\mathbb{Q}$ with conductor of a special shape are classified. The methods used in these papers are essentially non-diophantine methods. We also mention the list of elliptic curves over $\mathbb{Q}$ with conductor $N \leq 200$ in [Modular Functions of One Variable IV, Springer Lecture Notes 476(1975), pp. 81-113], which was found by a computer search. Some of the results mentioned in the present chapter may be summarized in the following table:

Number of elliptic curves over K in $E(S)$, where $S = S_T$, $T \subset \{2,3\}$, is the set of primes in $\mathcal{O}_K$ above T .

		K:		
T:		$\mathbb{Q}$	$\mathbb{Q}(i)$	$\mathbb{Q}(\sqrt{-2})$
	$\emptyset$	0	0	0
	$\{2\}$	24	64	40
	$\{3\}$	8	8	?
	$\{2,3\}$	760	1280(?)	?

Appendix

$$\text{Elliptic curves over } \mathbb{Q}(i)$$
$$\text{with conductor } N = (1+i)^a (3)^b$$

In the following tables we list the elliptic curves over $\mathbb{Q}(i)$ with conductor of the shape $N = (1+i)^a (3)^b$ for some $a,b \in \mathbb{N}_0$ (see 5.4).

Consider the diophantine equation

$$x^3 - y^2 = \xi(1+i)^{z_1} 3^{z_2}$$

with solutions (x,y,ξ,z_1,z_2) such that $x,y \in \mathbb{Z}[i]$, $\xi \in \{1,-1,i,-i\}$, $z_1,z_2 \in \mathbb{N}_0$. Let B be the set of all basic solutions (p,q,ε,m,n) of that equation satisfying the following additional property: $\mathrm{Re}(q) \geq 0$ and $\varepsilon \in \{1,i\}$, where $\mathrm{Im}(q) \geq 0$, if $\mathrm{Re}(q) = 0$. Then B is a set of representatives of non-associated basic solutions (for the definitions see 2.2). Let $(p,q,\varepsilon,m,n) \in B$ and let $\lambda = i^{\alpha}(1+i)^{\beta} 3^{\gamma}$ for some $0 \leq \alpha,\beta,\gamma \leq 1$. Then the equation

$$y^2 = x^3 - 3\lambda^2 px - 2\lambda^3 q$$

defines an elliptic curve over $\mathbb{Q}(i)$ with conductor $N = (1+i)^a (3)^b$ for some $a,b \in \mathbb{N}_0$ and conversely, each elliptic curve over $\mathbb{Q}(i)$ with conductor of that form arises uniquely in this way (see Theorem 2.4).

In table 1 the elements of B are listed. This list is complete up to a certain assumption (see 5.4). If (p,q,ε,m,n) is the j-th basic solution in the list and $\lambda = i^{\alpha}(1+i)^{\beta}3^{\gamma}$ for some $0 \leqq \alpha,\beta,\gamma \leqq 1$ then the elliptic curve with the above equation is denoted by $j*\lambda$.

In table 2 we list for each curve $j*\lambda$ the exponents (a,b) of the conductor at $1+i$ and 3 resp. and the isogeny graph associated to the isogeny class containing $j*\lambda$. The graphs are arranged in the table according to the lowest j occuring in each graph.

In table 3 the curves are sorted according to their conductors. The columns successively give

    (i)   The reference symbol $j*\lambda$ of the curve.

   (ii)   The coefficients a_1,a_2,a_3,a_4,a_6 of a global minimal Weierstrass equation of the curve.

 (iii)   The discriminant Δ of the global minimal Weierstrass equation, in factorized form.

  (iv)   The j-invariant of the curve.

   (v)   The Kodaira symbol for $1+i$ and 3 .

  (vi)   For each isogeny class of curves the associated isogeny graph.

In table 4 and 5 we give the number of curves and the number of isogeny classes resp. for any given conductor $N = (1+i)^{a}(3)^{b}$ with $a,b \in \mathbb{N}_0$.

Table 1

Basic solutions (p,q,ε,m,n) of the equation $x^3 - y^2 = \varepsilon(1+i)^m 3^n$

	(p,q)	p^3-q^2		(p,q)	p^3-q^2
1	(1,0)	1	30	(9-8i,19-36i)	$(1+i)^{13}$
2	(0,i)	1	31	(-9-8i,36-19i)	$i(1+i)^{13}$
3	(-2,3i)	1	32	(-17,71i)	$i(1+i)^{14}$
4	(-i,0)	i	33	(1+8i,9-20i)	$(1+i)^{15}$
5	(i,1-i)	i	34	(1-8i,9+20i)	$i(1+i)^{15}$
6	(23i,78-78i)	i	35	(-7,13)	$i(1+i)^{18}$
7	(-i,i)	1+i	36	(-1,2i)	3
8	(-i,1)	i(1+i)	37	(-i,1-i)	i3
9	(0,1-i)	$(1+i)^2$	38	(5i,8-8i)	i3
10	(1-i,1-i)	$i(1+i)^2$	39	(4079i,184211-184211i)	i3
11	(-3,5i)	$i(1+i)^2$	40	(i,1-2i)	(1+i)3
12	(-1,1)	$i(1+i)^2$	41	(1,2-i)	i(1+i)3
13	(1+i,1+i)	$i(1+i)^2$	42	(-2i,1+i)	$(1+i)^2 3$
14	(1+i,0)	$(1+i)^3$	43	(-14i,37+37i)	$(1+i)^2 3$
15	(-1+i,2)	$(1+i)^3$	44	(2,-3+i)	$(1+i)^2 3$
16	(1-2i,3i)	$(1+i)^3$	45	(-2,1-3i)	$(1+i)^2 3$
17	(1-i,0)	$i(1+i)^3$	46	(-3+2i,5+4i)	$(1+i)^2 3$
18	(-1-i,2)	$i(1+i)^3$	47	(3+2i,4+5i)	$(1+i)^2 3$
19	(1+2i,3i)	$i(1+i)^3$	48	(-1+3i,6-2i)	$(1+i)^3 3$
20	(-2,2i)	$(1+i)^4$	49	(-1-3i,6+2i)	$i(1+i)^3 3$
21	(-5,11i)	$(1+i)^4$	50	(-2,2)	$(1+i)^4 3$
22	(0,2)	$(1+i)^4$	51	(13,47)	$(1+i)^4 3$
23	(2-2i,2-4i)	$(1+i)^4$	52	(-1,5i)	$i(1+i)^6 3$
24	(2+2i,2+4i)	$(1+i)^4$	53	(-1,7i)	$(1+i)^8 3$
25	(-1,2+i)	$(1+i)^5$	54	(-7,3-16i)	$(1+i)^{11} 3$
26	(1,1+2i)	$i(1+i)^5$	55	(-7,3+16i)	$i(1+i)^{11} 3$
27	(-1,3i)	$i(1+i)^6$	56	(25,131)	$i(1+i)^{18} 3$
28	(-5-4i,14-9i)	$i(1+i)^8$	57	(23,11i)	$(1+i)^{24} 3$
29	(5-4i,9-14i)	$i(1+i)^8$			

	(p,q)	p^3-q^2		(p,q)	p^3-q^2
58	$(-505,23053i)$	$i(1+i)^{54}3$	96	$(-5i,7+7i)$	$i3^3$
59	$(2,i)$	3^2	97	$(-4-3i,9-8i)$	$(1+i)3^3$
60	$(0,3i)$	3^2	98	$(4-3i,8-9i)$	$i(1+i)3^3$
61	$(-3,6i)$	3^2	99	$(-6i,9+9i)$	$(1+i)^23^3$
62	$(-6,15i)$	3^2	100	$(3,9)$	$i(1+i)^23^3$
63	$(-40,253i)$	3^2	101	$(3+9i,9-27i)$	$i(1+i)^23^3$
64	$(-i,2-2i)$	$i3^2$	102	$(3-9i,9+27i)$	$i(1+i)^23^3$
65	$(-3i,3+3i)$	$i3^2$	103	$(-1+3i,9-i)$	$i(1+i)^23^3$
66	$(0,3-3i)$	$(1+i)^23^2$	104	$(-1-3i,9+i)$	$i(1+i)^23^3$
67	$(-3+3i,9+3i)$	$i(1+i)^23^2$	105	$(-7,17i)$	$i(1+i)^23^3$
68	$(-3-3i,9-3i)$	$i(1+i)^23^2$	106	$(3+3i,0)$	$(1+i)^33^3$
69	$(7,19)$	$i(1+i)^23^2$	107	$(-3-5i,16-2i)$	$(1+i)^33^3$
70	$(1-15i,37+45i)$	$i(1+i)^23^2$	108	$(3-3i,0)$	$i(1+i)^33^3$
71	$(1+15i,37-45i)$	$i(1+i)^23^2$	109	$(-3+5i,16+2i)$	$i(1+i)^33^3$
72	$(-3,3i)$	$i(1+i)^23^2$	110	$(-3,9)$	$(1+i)^43^3$
73	$(-3+i,2+2i)$	$(1+i)^33^2$	111	$(-2,10)$	$(1+i)^43^3$
74	$(-3-i,2-2i)$	$i(1+i)^33^2$	112	$(6,18)$	$(1+i)^43^3$
75	$(-3,3)$	$(1+i)^43^2$	113	$(366,7002)$	$(1+i)^43^3$
76	$(0,6)$	$(1+i)^43^2$	114	$(-15-16i,97-34i)$	$(1+i)^53^3$
77	$(4,10)$	$(1+i)^43^2$	115	$(15-16i,34-97i)$	$i(1+i)^53^3$
78	$(12,42)$	$(1+i)^43^2$	116	$(33,189)$	$i(1+i)^63^3$
79	$(-3,3+6i)$	$i(1+i)^43^2$	117	$(15,9i)$	$i(1+i)^{14}3^3$
80	$(3,6+3i)$	$i(1+i)^43^2$	118	$(-33,207i)$	$(1+i)^{16}3^3$
81	$(-3-12i,39+18i)$	$i(1+i)^63^2$	119	$(5745,435447)$	$i(1+i)^{26}3^3$
82	$(-3+12i,39-18i)$	$i(1+i)^63^2$	120	$(13,46)$	3^4
83	$(-3-4i,7+2i)$	$(1+i)^73^2$	121	$(0,9i)$	3^4
84	$(-3+4i,7-2i)$	$i(1+i)^73^2$	122	$(-9i,18+18i)$	$i3^4$
85	$(1,17)$	$i(1+i)^{10}3^2$	123	$(-11i,25+25i)$	$i3^4$
86	$(-9,21i)$	$i(1+i)^{10}3^2$	124	$(-27i,99+99i)$	$i3^4$
87	$(57,429)$	$i(1+i)^{14}3^2$	125	$(-12-13i,70-25i)$	$(1+i)3^4$
88	$(-105,1077i)$	$(1+i)^{16}3^2$	126	$(12-13i,25-70i)$	$i(1+i)3^4$
89	$(-33,177i)$	$i(1+i)^{18}3^2$	127	$(-4i,7-7i)$	$(1+i)^23^4$
90	$(-79+168i,2511-316i)$	$i(1+i)^{21}3^2$	128	$(-2+6i,17-9i)$	$(1+i)^23^4$
91	$(79+168i,316-2511i)$	$i(1+i)^{21}3^2$	129	$(2+6i,9-17i)$	$(1+i)^23^4$
92	$(15,183i)$	$(1+i)^{24}3^2$	130	$(0,9-9i)$	$(1+i)^23^4$
93	$(-73,827i)$	$i(1+i)^{30}3^2$	131	$(36i,153-153i)$	$(1+i)^23^4$
94	$(3,0)$	3^3	132	$(-9+18i,90-9i)$	$(1+i)^23^4$
95	$(-3i,0)$	$i3^3$	133	$(9+18i,9-90i)$	$(1+i)^23^4$
			134	$(-10,26i)$	$(1+i)^43^4$

	(p,q)	p^3-q^2
135	(0,18)	$(1+i)^4 3^3$
136	(-9,9i)	$i(1+i)^6 3^4$
137	(1809,76941)	$i(1+i)^6 3^4$
138	(-97,955i)	$i(1+i)^6 3^4$
139	(-9,45i)	$(1+i)^8 3^4$
140	(193,2681)	$(1+i)^8 3^4$
141	(-657,16839i)	$i(1+i)^{18} 3^4$
142	(153,963)	$i(1+i)^{30} 3^4$
143	(7,10)	3^5
144	(-i,11-11i)	$i3^5$
145	(-35i,146+146i)	$i3^5$
146	(3+7i,11-7i)	$i(1+i)^2 3^5$
147	(3-7i,11+7i)	$i(1+i)^2 3^5$
148	(-5,19)	$i(1+i)^2 3^5$
149	(-13,35i)	$(1+i)^4 3^5$
150	(3-16i,38+61i)	$i(1+i)^4 3^5$
151	(-3-16i,61+38i)	$i(1+i)^4 3^5$
152	(-1153,39151i)	$i(1+i)^{10} 3^5$
153	(-106,1090i)	$(1+i)^4 3^6$
154	(5i,34-34i)	$i3^7$
155	(23,73i)	$i(1+i)^6 3^7$
156	(73,595)	$(1+i)^8 3^7$
157	(-971i,21395+ 21395i)	$i3^8$
158	(19,215)	$i(1+i)^2 3^9$
159	(-47,2359)	$i(1+i)^{10} 3^{11}$
160	(239i,2761-2761i)	$i3^{13}$

Table 2

The exponents (a,b) of the conductors and the isogeny classes of the elliptic curves $j^*\lambda$.

$\lambda=$		1	i	1+i	i(1+i)	3	i3	(1+i)3	i(1+i)3
$1^*\lambda$		(8,2)	(6,2)	(10,2)	(10,2)	(8,2)	(6,2)	(10,2)	(10,2)
$63^*3\lambda$ $2^*3\lambda$ $2^*\lambda$ $63^*\lambda$		(0,3)	(8,3)	(10,3)	(10,3)	(0,3)	(8,3)	(10,3)	(10,3)
$3^*\lambda$		(6,3)	(8,3)	(10,3)	(10,3)	(6,3)	(8,3)	(10,3)	(10,3)
$4^*\lambda$		(12,2)	(12,2)	(12,2)	(12,2)	(12,2)	(12,2)	(12,2)	(12,2)
$5^*\lambda$		(12,3)	(12,3)	(12,3)	(12,3)	(12,3)	(12,3)	(12,3)	(12,3)
$6^*\lambda$		(12,3)	(12,3)	(12,3)	(12,3)	(12,3)	(12,3)	(12,3)	(12,3)
$7^*\lambda$		(13,3)	(13,3)	(13,3)	(13,3)	(13,3)	(13,3)	(13,3)	(13,3)
$8^*\lambda$		(13,3)	(13,3)	(13,3)	(13,3)	(13,3)	(13,3)	(13,3)	(13,3)
$9^*\lambda$ $9^*3\lambda$ $21^*i(1+i)\lambda$ $21^*i(1+i)3\lambda$		(10,2)	(10,2)	(2,2)	(8,2)	(10,2)	(10,2)	(2,2)	(8,2)
$10^*\lambda$		(13,3)	(13,3)	(13,3)	(13,3)	(13,3)	(13,3)	(13,3)	(13,3)

$\lambda=$		1	i	1+i	i(1+i)	3	i3	(1+i)3	i(1+i)3
$11^{*}\lambda$ $20^{*}i\lambda$	2	(10,2)	(10,2)	(9,2)	(9,2)	(10,2)	(10,2)	(9,2)	(9,2)
$12^{*}\lambda$		(10,3)	(10,3)	(9,3)	(9,3)	(10,3)	(10,3)	(9,3)	(9,3)
$13^{*}\lambda$		(13,3)	(13,3)	(13,3)	(13,3)	(13,3)	(13,3)	(13,3)	(13,3)
$14^{*}\lambda$		(14,2)	(14,2)	(14,2)	(14,2)	(14,2)	(14,2)	(14,2)	(14,2)
$15^{*}\lambda$		(14,3)	(14,3)	(14,3)	(14,3)	(14,3)	(14,3)	(14,3)	(14,3)
$16^{*}\lambda$		(11,3)	(11,3)	(11,3)	(11,3)	(11,3)	(11,3)	(11,3)	(11,3)
$17^{*}\lambda$		(14,2)	(14,2)	(14,2)	(14,2)	(14,2)	(14,2)	(14,2)	(14,2)
$18^{*}\lambda$		(14,3)	(14,3)	(14,3)	(14,3)	(14,3)	(14,3)	(14,3)	(14,3)
$19^{*}\lambda$		(11,3)	(11,3)	(11,3)	(11,3)	(11,3)	(11,3)	(11,3)	(11,3)
$22^{*}\lambda$ $22^{*}3\lambda$	3	(2,3)	(8,3)	(10,3)	(10,3)	(2,3)	(8,3)	(10,3)	(10,3)
$23^{*}\lambda$		(9,3)	(9,3)	(10,3)	(10,3)	(9,3)	(9,3)	(10,3)	(10,3)
$24^{*}\lambda$		(9,3)	(9,3)	(10,3)	(10,3)	(9,3)	(9,3)	(10,3)	(10,3)
$25^{*}\lambda$		(9,3)	(9,3)	(10,3)	(10,3)	(9,3)	(9,3)	(10,3)	(10,3)
$26^{*}\lambda$		(9,3)	(9,3)	(10,3)	(10,3)	(9,3)	(9,3)	(10,3)	(10,3)
$27^{*}\lambda$		(8,3)	(6,3)	(10,3)	(10,3)	(8,3)	(6,3)	(10,3)	(10,3)
$28^{*}i3\lambda$ $29^{*}\lambda$	3	(2,3)	(8,3)	(10,3)	(10,3)	(2,3)	(8,3)	(10,3)	(10,3)

$\lambda =$		1	i	1+i	i(1+i)	3	i3	(1+i)3	i(1+i)3
$30^*3\lambda$ $33^*\lambda$ $90^*\lambda$	(3, 3)	(1,3)	(8,3)	(10,3)	(10,3)	(1,3)	(8,3)	(10,3)	(10,3)
$31^*i3\lambda$ $34^*\lambda$ $91^*i\lambda$	(3, 3)	(1,3)	(8,3)	(10,3)	(10,3)	(1,3)	(8,3)	(10,3)	(10,3)
$32^*i3\lambda$ $35^*\lambda$ $93^*i\lambda$	(3, 3)	(1,3)	(8,3)	(10,3)	(10,3)	(1,3)	(8,3)	(10,3)	(10,3)
$36^*\lambda$	•	(6,4)	(8,4)	(10,4)	(10,4)	(6,4)	(8,4)	(10,4)	(10,4)
$37^*\lambda$ $39^*\lambda$	13	(12,4)	(12,4)	(12,4)	(12,4)	(12,4)	(12,4)	(12,4)	(12,4)
$38^*\lambda$	•	(12,4)	(12,4)	(12,4)	(12,4)	(12,4)	(12,4)	(12,4)	(12,4)
$40^*\lambda$	•	(13,4)	(13,4)	(13,4)	(13,4)	(13,4)	(13,4)	(13,4)	(13,4)
$41^*\lambda$	•	(13,4)	(13,4)	(13,4)	(13,4)	(13,4)	(13,4)	(13,4)	(13,4)
$42^*\lambda$	•	(10,4)	(10,4)	(8,4)	(3,4)	(10,4)	(10,4)	(8,4)	(3,4)
$43^*3\lambda$ $99^*\lambda$	3	(10,4)	(10,4)	(8,4)	(2,4)	(10,4)	(10,4)	(8,4)	(2,4)
$44^*\lambda$	•	(10,4)	(10,4)	(7,4)	(8,4)	(10,4)	(10,4)	(7,4)	(8,4)
$45^*\lambda$	•	(10,4)	(10,4)	(7,4)	(8,4)	(10,4)	(10,4)	(7,4)	(8,4)
$46^*\lambda$	•	(8,4)	(7,4)	(10,4)	(10,4)	(8,4)	(7,4)	(10,4)	(10,4)
$47^*\lambda$	•	(7,4)	(8,4)	(10,4)	(10,4)	(7,4)	(8,4)	(10,4)	(10,4)
$48^*\lambda$	•	(14,4)	(14,4)	(14,4)	(14,4)	(14,4)	(14,4)	(14,4)	(14,4)

λ=		1	i	1+i	i(1+i)	3	i3	(1+i)3	i(1+i)3
$49^*\lambda$	•	(14,4)	(14,4)	(14,4)	(14,4)	(14,4)	(14,4)	(14,4)	(14,4)
$50^*\lambda$	•	(10,4)	(10,4)	(9,4)	(9,4)	(10,4)	(10,4)	(9,4)	(9,4)
$51^*3\lambda$ $110^*\lambda$		(2,4)	(8,4)	(10,4)	(10,4)	(2,4)	(8,4)	(10,4)	(10,4)
$52^*\lambda$	•	(8,4)	(6,4)	(10,4)	(10,4)	(8,4)	(6,4)	(10,4)	(10,4)
$53^*\lambda$	•	(8,4)	(3,4)	(10,4)	(10,4)	(8,4)	(3,4)	(10,4)	(10,4)
$54^*\lambda$	•	(3,4)	(8,4)	(10,4)	(10,4)	(3,4)	(8,4)	(10,4)	(10,4)
$55^*\lambda$	•	(3,4)	(8,4)	(10,4)	(10,4)	(3,4)	(8,4)	(10,4)	(10,4)
$119^*\lambda$ $117^*i\lambda$ $58^*i3\lambda$ $56^*3\lambda$	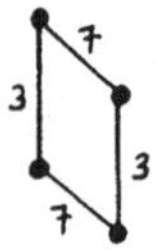	(1,4)	(8,4)	(10,4)	(10,4)	(1,4)	(8,4)	(10,4)	(10,4)
$57^*\lambda$ $118^*3\lambda$		(8,4)	(1,4)	(10,4)	(10,4)	(8,4)	(1,4)	(10,4)	(10,4)
$59^*\lambda$	•	(6,3)	(8,3)	(10,3)	(10,3)	(6,3)	(8,3)	(10,3)	(10,3)
$60^*\lambda$ $60^*3\lambda$		(0,5)	(8,5)	(10,5)	(10,5)	(0,5)	(8,5)	(10,5)	(10,5)
$61^*\lambda$	•	(8,5)	(6,5)	(10,5)	(10,5)	(8,5)	(6,5)	(10,5)	(10,5)
$62^*\lambda$	•	(6,5)	(8,5)	(10,5)	(10,5)	(6,5)	(8,5)	(10,5)	(10,5)
$64^*\lambda$	•	(12,3)	(12,3)	(12,3)	(12,3)	(12,3)	(12,3)	(12,3)	(12,3)
$65^*\lambda$	•	(12,5)	(12,5)	(12,5)	(12,5)	(12,5)	(12,5)	(12,5)	(12,5)
$66^*\lambda$ $66^*3\lambda$		(10,5)	(10,5)	(2,5)	(8,5)	(10,5)	(10,5)	(2,5)	(8,5)

$\lambda =$		1	i	1+i	i(1+i)	3	i3	(1+i)3	i(1+i)3
$67^{*}\lambda$	●	(13,5)	(13,5)	(13,5)	(13,5)	(13,5)	(13,5)	(13,5)	(13,5)
$68^{*}\lambda$	●	(13,5)	(13,5)	(13,5)	(13,5)	(13,5)	(13,5)	(13,5)	(13,5)
$69^{*}\lambda$	●	(10,3)	(10,3)	(9,3)	(9,3)	(10,3)	(10,3)	(9,3)	(9,3)
$70^{*}\lambda$	●	(13,3)	(13,3)	(13,3)	(13,3)	(13,3)	(13,3)	(13,3)	(13,3)
$71^{*}\lambda$	●	(13,3)	(13,3)	(13,3)	(13,3)	(13,3)	(13,3)	(13,3)	(13,3)
$72^{*}\lambda$	●	(10,5)	(10,5)	(9,5)	(9,5)	(10,5)	(10,5)	(9,5)	(9,5)
$73^{*}\lambda$	●	(14,3)	(14,3)	(14,3)	(14,3)	(14,3)	(14,3)	(14,3)	(14,3)
$74^{*}\lambda$	●	(14,3)	(14,3)	(14,3)	(14,3)	(14,3)	(14,3)	(14,3)	(14,3)
$75^{*}\lambda$	●	(3,5)	(8,5)	(10,5)	(10,5)	(3,5)	(8,5)	(10,5)	(10,5)
$76^{*}\lambda$ $76\ 3^{*}\lambda$	●●⎵3	(2,5)	(8,5)	(10,5)	(10,5)	(2,5)	(8,5)	(10,5)	(10,5)
$77^{*}\lambda$	●	(3,3)	(8,3)	(10,3)	(10,3)	(3,3)	(8,3)	(10,3)	(10,3)
$78^{*}\lambda$	●	(3,5)	(8,5)	(10,5)	(10,5)	(3,5)	(8,5)	(10,5)	(10,5)
$79^{*}\lambda$	●	(8,5)	(7,5)	(10,5)	(10,5)	(8,5)	(7,5)	(10,5)	(10,5)
$80^{*}\lambda$	●	(7,5)	(8,5)	(10,5)	(10,5)	(7,5)	(8,5)	(10,5)	(10,5)
$81^{*}\lambda$	●	(6,5)	(8,5)	(10,5)	(10,5)	(6,5)	(8,5)	(10,5)	(10,5)
$82^{*}\lambda$	●	(6,5)	(8,5)	(10,5)	(10,5)	(6,5)	(8,5)	(10,5)	(10,5)
$83^{*}\lambda$	●	(7,3)	(8,3)	(10,3)	(10,3)	(7,3)	(8,3)	(10,3)	(10,3)
$84^{*}\lambda$	●	(7,3)	(8,3)	(10,3)	(10,3)	(7,3)	(8,3)	(10,3)	(10,3)
$85^{*}\lambda$	●	(3,3)	(8,3)	(10,3)	(10,3)	(3,3)	(8,3)	(10,3)	(10,3)

$\lambda=$		1	i	1+i	i(1+i)	3	i3	(1+i)3	i(1+i)3
$86^*\lambda$		(8,5)	(3,5)	(10,5)	(10,5)	(8,5)	(3,5)	(10,5)	(10,5)
$87^*\lambda$ $89^*i3\lambda$	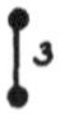 3	(1,5)	(8,5)	(10,5)	(10,5)	(1,5)	(8,5)	(10,5)	(10,5)
$88^*\lambda$ $92^*3\lambda$	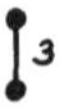 3	(8,5)	(1,5)	(10,5)	(10,5)	(8,5)	(1,5)	(10,5)	(10,5)
$94^*\lambda$ $116^*\lambda$	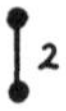 2	(6,2)	(8,2)	(10,2)	(10,2)	(6,0)	(8,0)	(10,0)	(10,0)
$95^*\lambda$		(12,2)	(12,2)	(12,2)	(12,2)	(12,0)	(12,0)	(12,0)	(12,0)
$96^*\lambda$ $96^*i\lambda$	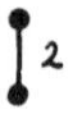 2	(12,2)	(12,2)	(12,2)	(12,2)	(12,0)	(12,0)	(12,0)	(12,0)
$97^*i\lambda$ $103^*\lambda$	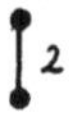 2	(13,2)	(13,2)	(13,2)	(13,2)	(13,0)	(13,0)	(13,0)	(13,0)
$98^*\lambda$ $104^*\lambda$	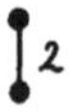 2	(13,2)	(13,2)	(13,2)	(13,2)	(13,0)	(13,0)	(13,0)	(13,0)
$100^*\lambda$		(10,4)	(10,4)	(9,4)	(9,4)	(10,4)	(10,4)	(9,4)	(9,4)
$101^*\lambda$		(13,4)	(13,4)	(13,4)	(13,4)	(13,4)	(13,4)	(13,4)	(13,4)
$102^*\lambda$		(13,4)	(13,4)	(13,4)	(13,4)	(13,4)	(13,4)	(13,4)	(13,4)
$105^*\lambda$ $114^*i(1+i)\lambda$ $111^*\lambda$ $115^*i(1+i)\lambda$	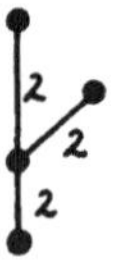 2, 2, 2	(10,2)	(10,2)	(9,2)	(9,2)	(10,0)	(10,0)	(9,0)	(9,0)
$106^*\lambda$		(14,2)	(14,2)	(14,2)	(14,2)	(14,0)	(14,0)	(14,0)	(14,0)
$107^*\lambda$ $109^*\lambda$	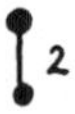 2	(14,2)	(14,2)	(14,2)	(14,2)	(14,0)	(14,0)	(14,0)	(14,0)

λ=		1	i	1+i	i(1+i)	3	i3	(1+i)3	i(1+i)3
$108^*\lambda$	•	(14,2)	(14,2)	(14,2)	(14,2)	(14,0)	(14,0)	(14,0)	(14,0)
$112^*\lambda$	•	(10,4)	(10,4)	(9,4)	(9,4)	(10,4)	(10,4)	(9,4)	(9,4)
$113^*\lambda$	•	(10,4)	(10,4)	(9,4)	(9,4)	(10,4)	(10,4)	(9,4)	(9,4)
$120^*\lambda$ $155^*\lambda$ $143^*i\lambda$ $138^*\lambda$		(8,2)	(6,2)	(10,2)	(10,2)	(8,1)	(6,1)	(10,1)	(10,1)
$121^*\lambda$ $121^*3\lambda$		(0,5)	(8,5)	(10,5)	(10,5)	(0,5)	(8,5)	(10,5)	(10,5)
$122^*\lambda$	•	(12,5)	(12,5)	(12,5)	(12,5)	(12,5)	(12,5)	(12,5)	(12,5)
$123^*\lambda$ $157^*\lambda$ $144^*\lambda$ $160^*\lambda$	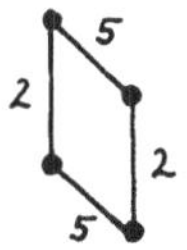	(12,2)	(12,2)	(12,2)	(12,2)	(12,1)	(12,1)	(12,1)	(12,1)
$124^*\lambda$	•	(12,5)	(12,5)	(12,5)	(12,5)	(12,5)	(12,5)	(12,5)	(12,5)
$125^*\lambda$ $146^*\lambda$		(13,2)	(13,2)	(13,2)	(13,2)	(13,1)	(13,1)	(13,1)	(13,1)
$126^*\lambda$ $147^*i\lambda$		(13,2)	(13,2)	(13,2)	(13,2)	(13,1)	(13,1)	(13,1)	(13,1)
$127^*(1+i)\lambda$ $140^*\lambda$ $149^*i\lambda$ $156^*\lambda$ $152^*i\lambda$ $159^*\lambda$	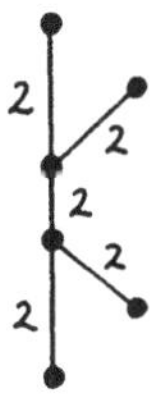	(3,2)	(8,2)	(10,2)	(10,2)	(3,1)	(8,1)	(10,1)	(10,1)
$128^*\lambda$ $151^*i(1+i)\lambda$		(10,2)	(10,2)	(7,2)	(8,2)	(10,1)	(10,1)	(7,1)	(8,1)

$\lambda =$		1	i	1+i	i(1+i)	3	i3	(1+i)3	i(1+i)3
$129^{*}\lambda$ $150^{*}(1+i)\lambda$	2	(10,2)	(10,2)	(7,2)	(8,2)	(10,1)	(10,1)	(7,1)	(8,1)
$130^{*}\lambda$ $130^{*}3\lambda$	3	(10,5)	(10,5)	(2,5)	(8,5)	(10,5)	(10,5)	(2,5)	(8,5)
$131^{*}\lambda$		(10,5)	(10,5)	(3,5)	(8,5)	(10,5)	(10,5)	(3,5)	(8,5)
$132^{*}\lambda$		(7,5)	(8,5)	(10,5)	(10,5)	(7,5)	(8,5)	(10,5)	(10,5)
$133^{*}\lambda$		(8,5)	(7,5)	(10,5)	(10,5)	(8,5)	(7,5)	(10,5)	(10,5)
$134^{*}\lambda$ $148^{*}\lambda$	2	(10,2)	(10,2)	(9,2)	(9,2)	(10,1)	(10,1)	(9,1)	(9,1)
$135^{*}\lambda$ $135^{*}3\lambda$	3	(2,5)	(8,5)	(10,5)	(10,5)	(2,5)	(8,5)	(10,5)	(10,5)
$136^{*}\lambda$		(8,5)	(6,5)	(10,5)	(10,5)	(8,5)	(6,5)	(10,5)	(10,5)
$137^{*}\lambda$		(6,5)	(8,5)	(10,5)	(10,5)	(6,5)	(8,5)	(10,5)	(10,5)
$139^{*}\lambda$		(8,5)	(3,5)	(10,5)	(10,5)	(8,5)	(3,5)	(10,5)	(10,5)
$141^{*}i3\lambda$ $142^{*}\lambda$	3	(1,5)	(8,5)	(10,5)	(10,5)	(1,5)	(8,5)	(10,5)	(10,5)
$145^{*}\lambda$ $154^{*}\lambda$	2	(12,2)	(12,2)	(12,2)	(12,2)	(12,1)	(12,1)	(12,1)	(12,1)
$153^{*}\lambda$ $158^{*}\lambda$	2	(10,2)	(10,2)	(9,2)	(9,2)	(10,1)	(10,1)	(9,1)	(9,1)

<u>Table 3</u>

Minimal equations, j-invariants, Kodaira reduction types
of the elliptic curves over $\mathbb{Q}(i)$ with conductor $N = (1+i)^a (3)^b$

$$\underline{N=(1+i)^6=(8)}$$

94^*3	O	O	O	-1	O	$-(1+i)^{12}$	$2^6 3^3$	I*2 ,IO	
116^*3	-1-i	i	-1-:	2-i	-i	$-i(1+i)^6$	$2^3 3^3 11^3$	II ,IO	2

$$\underline{N=(1+i)^8=(16)}$$

94^*i3	O	O	O	1	O	$(1+i)^{12}$	$2^6 3^3$	I*O ,IO	
116^*i3	O	O	O	11	14i	$i(1+i)^{18}$	$2^3 3^3 11^3$	I*6 ,IO	2

$$N=(1+i)^9=(16+16i)$$

$105^{*}(1+i)3$	O	$1-i$	O	$4i$	$4+4i$	$-i(1+i)^{20}$	$2^5 7^3$	I*7 ,IO
$114^{*}i3$	O	$1-i$	O	$-5-6i$	$-1+7i$	$(1+i)^{17}$	$(1+i)^7(2+3i)^3(6-i)^3$	I*4 ,IO
$111^{*}(1+i)3$	O	$1-i$	O	$-i$	O	$-(1+i)^{10}$	2^7	III ,IO
$115^{*}i3$	O	$1-i$	O	$5-6i$	$7-i$	$i(1+i)^{17}$	$i(1+i)^7(2-3i)^3(6+i)^3$	II* ,IO
$105^{*}i(1+i)3$	O	$1+i$	O	$-4i$	$4-4i$	$i(1+i)^{20}$	$2^5 7^3$	I*7 ,IO
$114^{*}3$	O	$-1-i$	O	$5+6i$	$-7-i$	$-(1+i)^{17}$	$(1+i)^7(2+3i)^3(6-i)^3$	II* ,IO
$111^{*}i(1+i)3$	O	$1+i$	O	i	O	$(1+i)^{10}$	2^7	III ,IO
$115^{*}3$	O	$-1-i$	O.	$-5+6i$	$1+7i$	$-i(1+i)^{17}$	$i(1+i)^7(2-3i)^3(6+i)^3$	I*4 ,IO

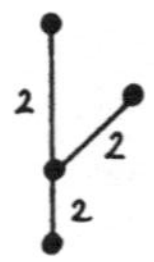

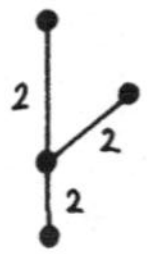

$$N=(1+i)^{10}=(32)$$

$94^{*}(1+i)3$	O	O	O	$-2i$	O	$-(1+i)^{18}$	$2^6 3^3$	I*4 ,IO
$116^{*}(1+i)3$	O	O	O	$-22i$	$28-28i$	$-i(1+i)^{24}$	66^3	I*10,IO
$94^{*}i(1+i)3$	O	O	O	$2i$	O	$(1+i)^{18}$	$2^6 3^3$	I*4 ,IO
$116^{*}i(1+i)3$	O	O	O	$22i$	$-28-28i$	$i(1+i)^{24}$	66^3	I*10,IO
$105^{*}3$	O	$-i$	O	2	$-2i$	$-i(1+i)^{14}$	$2^5 7^3$	I*O ,IO
$114^{*}i(1+i)3$	O	-1	O	$11-10i$	$-23-6i$	$(1+i)^{23}$	$(1+i)^7(2+3i)^3(6-i)^3$	I*9 ,IO
$111^{*}3$	O	-1	O	1	-1	$-(1+i)^{16}$	2^7	I*2 ,IO
$115^{*}i(1+i)3$	O	-1	O	$11+10i$	$-23+6i$	$i(1+i)^{23}$	$i(1+i)^7(2-3i)^3(6+i)^3$	I*9 ,IO

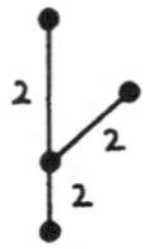

105^*i3	O	1	O	-2	-2	$i(1+i)^{14}$	$2^5 7^3$	I*O ,IO
$114^*(1+i)3$	O	i	O	$-11+10i$	$6-23i$	$-(1+i)^{23}$	$(1+i)^7(2+3i)^3(6-i)^3$	I*9 ,IO
111^*i3	O	$-i$	O	-1	i	$(1+i)^{16}$	2^7	I*2 ,IO
$115^*(1+i)$	O	i	O	$-11-10i$	$-6-23i$	$-i(1+i)^{23}$	$i(1+i)^7(2-3i)^3(6+i)^3$	I*9 ,IO

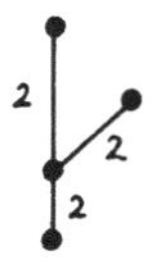

$$N=(1+i)^{12}=(64)$$

95^*3	O	O	O	i	O	$-i(1+i)^{12}$	12^3	II ,IO
95^*i3	O	O	O	$-i$	O	$i(1+i)^{12}$	12^3	II ,IO
$95^*(1+i)3$	O	O	O	-2	O	$-i(1+i)^{18}$	12^3	I*2 ,IO
$95^*i(1+i)3$	O	O	O	2	O	$i(1+i)^{18}$	12^3	I*2 ,IO
96^*3	O	$-1+i$	O	i	$-1-i$	$-i(1+i)^{12}$	20^3	II ,IO 
96^*i3	O	$-1-i$	O	$-i$	$-1+i$	$i(1+i)^{12}$	20^3	II ,IO
$96^*(1+i)3$	O	1	O	-3	1	$-i(1+i)^{18}$	20^3	I*2 ,IO 
$96^*i(1+i)3$	O	i	O	3	$-i$	$i(1+i)^{18}$	20^3	I*2 ,IO

$$N=(1+i)^{13}=(64+64i)$$

$103^{*}3$	O	$-i$	O	$-i$	-1	$-i(1+i)^{14}$ $(1+i)^{13}(2-i)^{3}$	III , IO	
$97^{*}i3$	O	-1	O	$-1-i$	$1+i$	$(1+i)^{13}$ $-i(1+i)^{11}(2-i)^{6}$	II , IO	
$103^{*}i3$	O	1	O	i	i	$i(1+i)^{14}$ $(1+i)^{13}(2-i)^{3}$	III , IO	
$97^{*}3$	O	i	O	$1+i$	$-1+i$	$-(1+i)^{13}$ $-i(1+i)^{11}(2-i)^{6}$	II , IO	
$103^{*}(1+i)3$	O	$1-i$	O	2	$2-2i$	$-i(1+i)^{20}$ $(1+i)^{13}(2-i)^{3}$	III*, IO	
$97^{*}i(1+i)3$	O	$-1-i$	O	$2-2i$	-4	$-(1+i)^{19}$ $-i(1+i)^{11}(2-i)^{6}$	I 2 , IO	
$103^{*}i(1+i)3$	O	$1+i$	O	-2	$-2-2i$	$i(1+i)^{20}$ $(1+i)^{13}(2-i)^{3}$	III*, IO	
$97^{*}(1+i)3$	O	$-1+i$	O	$-2+2i$	$-4i$	$(1+i)^{19}$ $-i(1+i)^{11}(2-i)^{6}$	I*2 , IO	
$104^{*}3$	O	i	O	i	-1	$-i(1+i)^{14}$ $-i(1+i)^{13}(2+i)^{3}$	III , IO	
$98^{*}3$	O	1	O	$-1+i$	$-1+i$	$-i(1+i)^{13}$ $-(1+i)^{11}(2+i)^{6}$	II , IO	
$104^{*}i3$	O	-1	O	$-i$	i	$i(1+i)^{14}$ $-i(1+i)^{13}(2+i)^{3}$	III , IO	
$98^{*}i3$	O	i	O	$1-i$	$1+i$	$i(1+i)^{13}$ $-(1+i)^{11}(2+i)^{6}$	II , IO	
$104^{*}(1+i)3$	O	$-1+i$	O	-2	$2-2i$	$-i(1+i)^{20}$ $-i(1+i)^{13}(2+i)^{13}$	III*, IO	
$98^{*}(1+i)3$	O	$1+i$	O	$-2-2i$	$-4i$	$-i(1+i)^{19}$ $-(1+i)^{11}(2+i)^{6}$	I*2 , IO	
$104^{*}i(1+i)3$	O	$-1-i$	O	2	$-2-2i$	$i(1+i)^{20}$ $-i(1+i)^{13}(2+i)^{13}$	III*, IO	
$98^{*}i(1+i)3$	O	$-1+i$	O	$2+2i$	-4	$i(1+i)^{19}$ $-(1+i)^{11}(2+i)^{6}$	I*2 , IO	

$$N=(1+i)^{14}=(128)$$

106^*3	o	o	o	$-1-i$	o	$-(1+i)^{15}$	12^3	III ,IO	•
106^*i3	o	o	o	$1+i$	o	$(1+i)^{15}$	12^3	III ,IO	•
$106^*(1+i)3$	o	o	o	$2-2i$	o	$-(1+i)^{21}$	12^3	III*,IO	•
$106^*i(1+i)3$	o	o	o	$-2+2i$	o	$(1+i)^{21}$	12^3	III*,IO	•
108^*3	o	o	o	$-1+i$	o	$-i(1+i)^{15}$	12^3	III ,IO	•
108^*i3	o	o	o	$1-i$	o	$i(1+i)^{15}$	12^3	III ,IO	•
$108^*(1+i)3$	o	o	o	$-2-2i$	o	$-i(1+i)^{21}$	12^3	III*,IO	•
$108^*i(1+i)3$	o	o	o	$2+2i$	o	$i(1+i)^{21}$	12^3	III*,IO	•
107^*3	o	$-1+i$	o	$1+i$	-2	$-(1+i)^{15}$	$-2^6(4+i)^3$	III ,IO	
109^*3	o	$-1-i$	o	$1-i$	-2	$-i(1+i)^{15}$	$-2^6(4-i)^3$	III ,IO	
107^*i3	o	$-1-i$	o	$-1-i$	$2i$	$(1+i)^{15}$	$-2^6(4+i)^3$	III ,IO	
109^*i3	o	$1-i$	o	$-1+i$	$2i$	$i(1+i)^{15}$	$-2^6(4-i)^3$	III ,IO	
$107^*(1+i)3$	o	1	o	$-3+2i$	$1-2i$	$-(1+i)^{21}$	$-2^6(4+i)^3$	III*,IO	
$109^*(1+i)3$	o	i	o	$3+2i$	$2-i$	$-i(1+i)^{21}$	$-2^6(4-i)^3$	III*,IO	
$107^*i(1+i)3$	o	i	o	$3-2i$	$-2-i$	$(1+i)^{21}$	$-2^6(4+i)^3$	III*,IO	
$109^*i(1+i)3$	o	-1	o	$-3-2i$	$-1-2i$	$i(1+i)^{21}$	$-2^6(4-i)^3$	III*,IO	

$$N=(1+i)^3(3)=(6+6i)$$

127*(1+i)3	O	-1	O	1	O	$-(1+i)^8 3$	$2^{11}/3$	I*1 ,I1	
140*3	-1-i	-i	-1-i	16-i	-28i	$-(1+i)^8 3$	$2^2 193^3/3$	I*1 ,I1	
149*i3	-1-i	-i	-1-i	1-i	-i	$(1+i)^4 3^2$	$2^4 13^3/3^2$	III ,I2	
156*3	-1-i	O	-1-i	6-i	-5i	$-(1+i)^8 3^4$	$2^2 73^3/3^4$	I*1 ,I4	
152*i3	-1-i	O	-1-i	96-i	-347i	$i(1+i)^{10} 3^2$	$2\ 1153^3/3^2$	III*,I2	
159*3	-1-i	O	-1-i	-4-i	-23i	$-i(1+i)^{10} 3^8$	$2\ 47^3/3^8$	III*,I8	

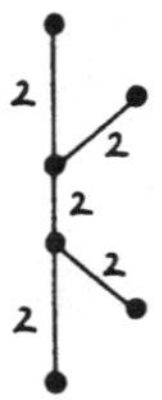

$$N=(1+i)^6(3)=(24)$$

120*i3	O	-i	O	4	2i	$(1+i)^{12} 3$	$2^6 13^3/3$	I*2 ,I1	
155*i3	-1-i	O	O	-2	-i	$i(1+i)^6 3^4$	$2^3 23^3/3^4$	II ,I4	
143*3	O	-1	O	-2	O	$-(1+i)^{12} 3^2$	$2^6 7^3/3^2$	I*2 ,I2	
138*i3	-1-i	O	-1-i	8-i	-8i	$-i(1+i)^6 3$	$2^3 97^3/3$	II ,I1	

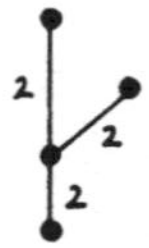

$$N=(1+i)^7(3)=(24+24i)$$

128*(1+i)3	O	-1+i	O	-1-i	1	$-(1+i)^8 3$	$-(1+i)^{19}(2-i)^3/3$	III ,I1	
151*i3	O	1-i	O	-1-6i	-5+3i	$i(1+i)^{16} 3^2$	$-2^4(2-i)^3(7+2i)^3/3^2$	I*5 ,I2	
129*(1+i)3	O	-1-i	O	-1+i	1	$-(1+i)^8 3$	$-i(1+i)^{19}(2+i)^3/3$	III ,I1	
150*3	O	1+i	O	-1+6i	-5-3i	$-i(1+i)^{16} 3^2$	$-2^4(2+i)^3(7-2i)^3/3^2$	I*5 ,I2	

$$N = (1+i)^8 (3) = (48)$$

128 *i(1+i)3	O	-1-i	O	1+i	-i	$(1+i)^8 3$		II	 , I1
151 *3	O	-1-i	O	1+6i	-3-5i	$-i(1+i)^{16} 3^2$		II*	, I2
129 *i(1+i)3	O	1-i	O	1-i	-i	$(1+i)^8 3$		II	 , I1
150 *i3	O	-1+i	O	1-6i	-3+5i	$i(1+i)^{16} 3^2$		II*	, I2
120 *3	O	-1	O	-4	-2	$-(1+i)^{12} 3$	$2^6 13^3/3$	I*0	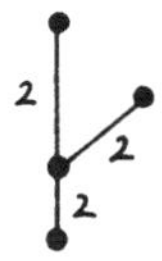 , I1
155 *3	O	i	O	-8	-8i	$-i(1+i)^{18} 3^4$	$2^3 23^3/3^4$	I*6	, I4
143 *i3	O	-i	O	2	O	$(1+i)^{12} 3^2$	$2^6 7^3/3^2$	I*0	, I2
138 *3	O	i	O	32	-60i	$i(1+i)^{18} 3$	$2^3 97^3/3$	I*6	, I1
127 *i(1+i)3	O	-i	C	-1	O	$(1+i)^8 3$	$2^{11}/3$	II	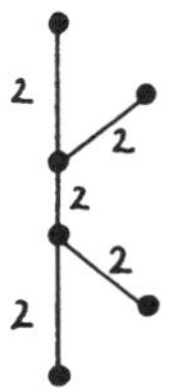 , I1
140 *i3	O	i	C	64	220i	$(1+i)^{20} 3$	$2^2 193^3/3$	I*8	, I1
149 *3	O	-i	C	4	-4i	$-(1+i)^{16} 3^2$	$2^4 13^3/3^2$	I*4	, I2
156 *i3	O	-i	C	24	36i	$(1+i)^{20} 3^4$	$2^2 73^3/3^4$	I*8	, I4
159 *i3	O	-i	C	-16	180i	$i(1+i)^{22} 3^8$	$2\ 47^3/3^8$	I*10	, I8
152 *3	O	i	C	384	-2772i	$-i(1+i)^{22} 3^2$	$2\ 1153^3/3^2$	I*10	, I2

$$N = (1+i)^9 (3) = (48+48i)$$

148 *(1+i)3	O	-1-i	O	4i	4-4i	$-i(1+i)^{20} 3^2$	$2^5 5^3/3^2$	I*7	 , I2
134 *(1+i)3	O	1+i	O	-i	O	$-(1+i)^{10} 3$	$2^7 5^3/3$	III	, I1

$148^{*}i(1+i)3$	O	$1-i$	O	$-4i$	$-4-4i$	$i(1+i)^{20}3^{2}$	$2^{5}5^{3}/3^{2}$	I*7 , I2	•—•²
$134^{*}i(1+i)3$	O	$-1+i$	O	i	O	$(1+i)^{10}3$	$2^{7}5^{3}/3$	III , I1	
$158^{*}(1+i)3$	O	$1+i$	O	$-12i$	$36-36i$	$-i(1+i)^{20}3^{6}$	$-2^{5}19^{3}/3^{6}$	I*7 , I6	•—•²
$153^{*}(1+i)3$	O	$-1-i$	O	$-17i$	$-26+26i$	$-(1+i)^{10}3^{3}$	$2^{7}53^{3}/3^{3}$	III , I3	
$158^{*}i(1+i)3$	O	$-1+i$	O	$12i$	$-36-36i$	$i(1+i)^{20}3^{6}$	$-2^{5}19^{3}/3^{6}$	I*7 , I6	•—•²
$153^{*}i(1+i)3$	O	$1-i$	O	$17i$	$26+26i$	$(1+i)^{10}3^{3}$	$2^{7}53^{3}/3^{3}$	III , I3	

$$N=(1+i)^{10}(3)=(96)$$

$128^{*}3$	O	1	O	$1-2i$	-1	$-(1+i)^{14}3$	$-(1+i)^{19}(2-i)^{3}/3$	I*O , I1	•—•²
$151^{*}i(1+i)3$	O	-1	O	$11-2i$	$-7-$	$i(1+i)^{22}3^{2}$	$-2^{4}(2-i)^{3}(7+2i)^{3}/3^{2}$	I*8 , I2	
$128^{*}i3$	O	i	O	$-1+2i$	i	$(1+i)^{14}3$	$-(1+i)^{19}(2-i)^{3}/3$	I*O , I2	•—•²
$151^{*}(1+i)3$	O	i	O	$-11+2i$	$14-7i$	$-i(1+i)^{22}3^{2}$	$-2^{4}(2-i)^{3}(7+2i)^{2}/3^{2}$	I*8 , I2	
$129^{*}3$	O	i	O	$-1-2i$	i	$-(1+i)^{14}3$	$-i(1+i)^{19}(2+i)^{3}/3$	I*O , I1	•—•²
$150^{*}(1+i)3$	O	$-i$	O	$-11-2i$	$14+7i$	$-i(1+i)^{22}3^{2}$	$-2^{4}(2+i)^{3}(7-2i)^{3}/3^{2}$	I*8 , I2	
$129^{*}i3$	O	-1	O	$1+2i$	1	$(1+i)^{14}3$	$-i(1+i)^{19}(2+i)^{3}/3$	I*O , I1	•—•²
$150^{*}i(1+i)3$	O	1	O	$11+2i$	$7-14i$	$i(1+i)^{22}3^{2}$	$-2^{4}(2+i)^{3}(7-2i)^{3}/3^{2}$	I*8 , I2	

Label								I-pair	Diagram
148*3	O	-1	O	2	-2	$-i(1+i)^{14}3^2$	$2^5 5^3/3^2$	I*O , I2	•—• (2)
134*3	O	-i	O	3	-3i	$-(1+i)^{16}3$	$2^7 5^3/3$	I*2 , I1	
148*i3	O	-i	O	-2	2i	$i(1+i)^{14}3^2$	$2^5 5^3/3^2$	I*O , I2	•—• (2)
134*i3	O	1	O	-3	-3	$(1+i)^{16}3$	$2^7 5^3/3$	I*2 , I1	
158*3	O	1	O	-6	-18	$-i(1+i)^{14}3^6$	$-2^5 19^3/3^6$	I*O , I6	•—• (2)
153*3	O	i	O	35	-69i	$-(1+i)^{16}3^3$	$2^7 53^3/3^3$	I*2 , I3	
158*i3	O	i	O	6	18i	$i(1+i)^{14}3^6$	$-2^5 19^3/3^6$	I*O , I6	•—• (2)
153*i3	O	-1	O	-35	-69	$(1+i)^{16}3^3$	$2^7 53^3/3^3$	I*2 , I3	
120*(1+i)3	O	-1-i	O	-8i	4-4i	$-(1+i)^{18}3$	$2^6 13^3/3$	I*4 , I1	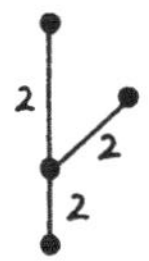 •—•—• with branch •, edges (2)
155*(1+i)3	O	-1+i	O	-16i	16+16i	$-i(1+i)^{24}3^4$	$2^3 23^3/3^4$	I*10, I4	
143*i(1+i)3	O	1-i	O	4i	O	$(1+i)^{18}3^2$	$2^6 7^3/3^2$	I*4 , I2	
138*(1+i)3	O	-1+i	O	64i	120+120i	$i(1+i)^{24}3$	$2^3 97^3/3$	I*10, I1	
120*i(1+i)3	O	1-i	O	8i	-4-4i	$(1+i)^{18}3$	$2^6 13^3/3$	I*4 , I1	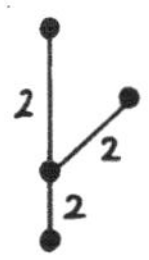 •—•—• with branch •, edges (2)
155*i(1+i)3	O	-1-i	O	16i	16-16i	$i(1+i)^{24}3^4$	$2^3 23^3/3^4$	I*10, I4	
143*(1+i)3	O	-1-i	O	-4i	O	$-(1+i)^{18}3^2$	$2^6 7^3/3^2$	I*4 , I2	
138*i(1+i)3	O	-1-i	O	-64i	120-120i	$-i(1+i)^{24}3$	$2^3 97^3/3$	I*10, I1	
127*3	O	-1-i	O	2i	O	$-(1+i)^{14}3$	$2^{11}/3$	I*O , I1	larger weighted tree, edges (2)
140*(1+i)3	O	1+i	O	-128i	440-440i	$-(1+i)^{26}3$	$2^2 193^3/3$	I*12, I1	
149*i(1+i)3	O	1+i	O	-8i	8-8i	$(1+i)^{22}3^2$	$2^4 13^3/3^2$	I*8 , I2	
156*(1+i)3	O	-1-i	O	-48i	72-72i	$-(1+i)^{26}3^4$	$2^2 73^3/3^4$	I*12, I4	
159*(1+i)3	O	-1-i	O	32i	360-360i	$-i(1+i)^{28}3^8$	$2\,47^3/3^8$	I*14, I8	
152*i(1+i)3	O	-1-i	O	-768i	5544-5544i	$i(1+i)^{28}3^2$	$2\,1153^3/3^2$	I*14, I2	

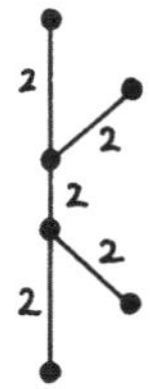

- 116 -

127 *i3	O	1-i	O	-2i	O	$(1+i)^{14}3$	$2^{11}/3$	I*O , I1
140 *i(1+i)3	O	-1+i	O	128i	-440-440i	$(1+i)^{26}3$	$2^2193^3/3$	I*12, I1
149 *(1+i)3	O	1-i	O	8i	8+8i	$-(1+i)^{22}3^2$	$2^413^3/3^2$	I*8 , I2
156 *i(1+i)3	O	1-i	O	48i	-72-72i	$(1+i)^{26}3^4$	$2^273^3/3^4$	I*12, I4
159 *i(1+i)3	O	1-i	O	-32i	-360-360i	$i(1+i)^{28}3^8$	$2\ 47^3/3^8$	I*14, I8
152 *(1+i)3	O	-1+i	O	768i	5544+5544i	$-i(1+i)^{28}3^2$	$2\ 1153^3/3^2$	I*14, I2

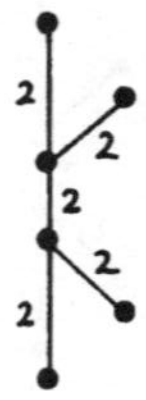

$$N=(1+i)^{12}(3)=(192)$$

154 *3	O	-1-i	O	-i	-3+3i	$-i(1+i)^{12}3^4$	$-2^65^3/3^4$	II , I4 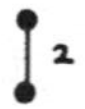
145 *3	O	1-i	O	11i	-7-7i	$-i(1+i)^{12}3^2$	$2^65^37^3/3^2$	II , I2
154 *(1+i)3	O	i	O	3	-9i	$-i(1+i)^{18}3^4$	$-2^65^3/3^4$	I*2 , I4 
145 *(1+i)3	O	-1	O	-23	51	$-i(1+i)^{18}3^2$	$2^65^37^3/3^2$	I*2 , I2
154 *i3	O	1-i	O	i	3+3i	$i(1+i)^{12}3^4$	$-2^65^3/3^4$	II , I4 
145 *i3	O	1+i	O	-11i	-7+7i	$i(1+i)^{12}3^2$	$2^65^37^3/3^2$	II , I2
154 *i(1+i)3	O	-1	O	-3	-9	$i(1+i)^{18}3^4$	$-2^65^3/3^4$	I*2 , I4 
145 *i(1+i)3	O	-i	O	23	-51i	$i(1+i)^{18}3^2$	$2^65^37^3/3^2$	I*2 , I2

123*3	o	$-1+i$	o	$3i$	$-3-3i$	$-i(1+i)^{12}3^1$ $2^6 11^3/3$	II , I1
157*3	o	$1-i$	o	$323i$	$-1477-1477i$	$-i(1+i)^{12}3^5$ $2^6 971^3/3^5$	II , I5
144*3	o	$1+i$	o	i	$-1+i$	$-i(1+i)^{12}3^2$ $2^6/3^2$	II , I2
160*3	o	$-1-i$	o	$-79i$	$-231+231i$	$-i(1+i)^{12}3^{10}$ $-2^6 239^3/3^{10}$	II , I10
123*i3	o	$-1-i$	o	$-3i$	$-3+3i$	$i(1+i)^{12}3$ $2^6 11^3/3$	II , I1
157*i3	o	$1+i$	o	$-323i$	$-1477+1477i$	$i(1+i)^{12}3^5$ $2^6 971^3/3^5$	II , I5
144*i3	o	$-1+i$	o	$-i$	$1+i$	$i(1+i)^{12}3^2$ $2^6/3^2$	II , I2
160*i3	o	$1-i$	o	$79i$	$231+231i$	$i(1+i)^{12}3^{10}$ $-2^6 239^3/3^{10}$	II , I10
123*(1+i)3	o	1	o	-7	5	$-i(1+i)^{18}3$ $2^6 11^3/3$	I*2 , I1
157*(1+i)3	o	-1	o	-647	6555	$-i(1+i)^{18}3^5$ $2^6 971^3/3^5$	I*2 , I5
144*(1+i)3	o	$-i$	o	-1	$-3i$	$-i(1+i)^{18}3^2$ $2^6/3^2$	I*2 , I2
160*(1+i)3	o	i	o	159	$-765i$	$-i(1+i)^{18}3^{10}$ $-2^6 239^3/3^{10}$	I*2 I10
123*i(1+i)3	o	i	o	7	$-5i$	$i(1+i)^{18}3$ $2^6 11^3/3$	I*2 , I1
157*i(1+i)3	o	$-i$	o	647	$-6555i$	$i(1+i)^{18}3^5$ $2^6 971^3/3^5$	I*2 , I5
144*i(1+i)3	o	1	o	1	-3	$i(1+i)^{18}3^2$ $2^6/3^2$	I*2 , I2
160*i(1+i)3	o	$-i$	o	-159	-765	$i(1+i)^{18}3^{10}$ $-2^6 239^3/3^{10}$	I*2 , I10

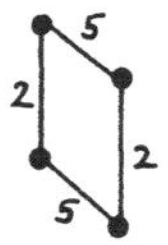

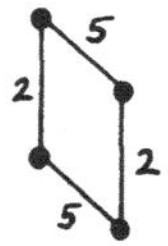

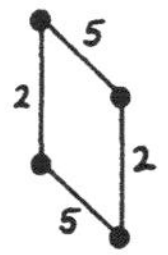

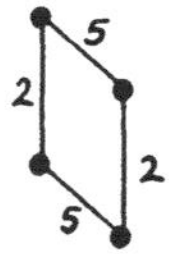

$$N=(1+i)^{13}(3)=(192+192i)$$

146*3	o	$1-i$	o	$-1-3i$	-2	$-i(1+i)^{14}3^2$ $i(1+i)^{13}(5+2i)^3/3^2$	III , I2
125*3	o	$-1-i$	o	$4+3i$	$-5-i$	$-(1+i)^{13}3$ $-i(1+i)^{14}(13-12i)^3/3$	II , I1

146*i3	o	1+i	o	1+3i	2i	$i(1+i)^{14}3^2$	$i(1+i)^{13}(5+2i)^3/3^2$	III , I2
125*i3	o	1-i	o	-4-5i	-1+5i	$(1+i)^{13}3$	$(1+i)^{11}(12+13i)^3/3$	II , I1
146*(1+i)3	o	-1	o	5-2i	-1-2i	$-i(1+i)^{20}3^2$	$i(1+i)^{13}(5+2i)^3/3^2$	III*, I2
125*(1+i)3	o	i	o	-9+8i	4-17i	$-(1+i)^{19}3$	$(1+i)^{11}(12+13i)^3/3$	I*2 , I1
146*i(1+i)3	o	-i	o	-5+2i	-2+i	$i(1+i)^{20}3^2$	$i(1+i)^{13}(5+2i)^3/3^2$	III*, I2
125*i(1+i)3	o	-1	o	9+8i	-17-4i	$i(1+i)^{19}3$	$(1+i)^{11}(12+13i)^3/3$	I*2 , I1
147*3	o	1+i	o	-1+3i	-2	$-i(1+i)^{14}3^2$	$-(1+i)^{13}(5-2i)^3/3^2$	III , I2
126*i3	o	1-i	o	4-5i	5-i	$i(1+i)^{13}3$	$-(1+i)^{11}(13+12i)^3/3$	II , I1
147*i3	o	-1+i	o	1-3i	2i	$i(1+i)^{14}3^2$	$-(1+i)^{13}(5-2i)^3/3^2$	III , I2
126*3	o	-1-i	o	-4+5i	1+5i	$-i(1+i)^{13}3$	$-(1+i)^{11}(13+12i)^3/3$	II , I1
147*(1+i)3	o	-i	o	-5-2i	2+i	$-i(1+i)^{20}3^2$	$-(1+i)^{13}(5-2i)^3/3^2$	III*, I2
126*i(1+i)3	o	-1	o	9+8i	-17+4i	$i(1+i)^{19}3$	$-(1+i)^{11}(13+12i)^3/3$	I*2 , I1
147*i(1+i)3	o	1	o	5+2i	1-2i	$i(1+i)^{20}3^2$	$-(1+i)^{13}(5-2i)^3/3^2$	III*, I2
126*(1+i)3	o	i	o	-9-8i	-4-17i	$-i(1+i)^{19}3$	$-(1+i)^{11}(13+2i)^3/3$	I*2 , I1

$$N=(1+i)^2(3)^2=(18)$$

9*(1+i)3	O	O	O	O	27	$-(1+i)^8 3^9$	O	IV* , III*
9*(1+i)	O	O	O	O	1	$-(1+i)^8 3^3$	O	IV* , III
21*i3	-1-i	i	-1-i	33-i	-58i	$-(1+i)^4 3^9$	$2^4 3^3 5^3$	IV , III*
21*i	-1-i	i	O	3	-i	$-(1+i)^4 3^3$	$2^4 3^3 5^3$	IV , III

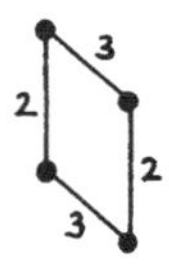

$$N=(1+i)^3(3)^2=(18+18i)$$

127*1+i	O	O	O	6	7	$-(1+i)^8 3^7$	$2^{11}/3$	I*1 , I*1
140*1	-1-i	i	O	144	-598i	$-(1+i)^8 3^7$	$2^2 193^3/3$	I*1 , I*1
149*i	-1-i	i	O	9	-4i	$(1+i)^4 3^8$	$2^4 13^3/3^2$	III , I*2
156*1	-1-i	i	-1-i	54-i	-122i	$-(1+i)^8 3^{10}$	$2^2 73^3/3^4$	I*1 , I*4
152*i	-1-i	i	-1-i	864-i	-9356i	$i(1+i)^{10} 3^8$	$2 \cdot 1153^3/3^2$	III*, I*2
159*1	-1-i	i	-1-i	-36-i	-608i	$-i(1+i)^{10} 3^{14}$	$2 \cdot 47^3/3^8$	III*, I*8

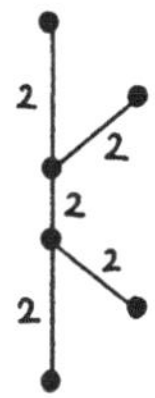

$$N=(1+i)^6(3)^2=(72)$$

1*i3	O	O	O	27	O	$(1+i)^{12} 3^9$	$2^6 3^3$	I*2 , III*
1*i	O	O	O	3	O	$(1+i)^{12} 3^3$	$2^6 3^3$	I*2 , III

94*1	O	O	O	-9	O	$-(1+i)^{12}3^6$	$2^6 3^3$	I*2 , I*O
116*1	-1-i	i	O	24	-35i	$-i(1+i)^6 3^6$	$2^3 3^3 11^3$	II , I*O
120*i	O	O	O	39	92i	$(1+i)^{12}3^7$	$2^6 13^3/3$	I*2 , I*1 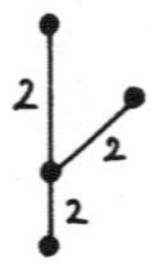
155*i	-1-i	i	O	-18	-27i	$i(1+i)^6 3^{10}$	$2^3 23^3/3^4$	II , I*4
143*1	O	O	O	-21	-20	$-(1+i)^{12}3^8$	$2^6 7^3/3^2$	I*2 , I*2
138*i	-1-i	i	-1-i	72-i	-203i	$-i(1+i)^6 3^7$	$2^3 97^3/3$	II , I*1

$$N=(1+i)^7(3)^2=(72+72i)$$

128*1+i	O	O	O	-9-3i	13+4i	$-(1+i)^8 3^7$	$-(1+i)^{19}(2-i)^3/3$	III , I*1 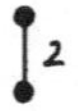
151*i	O	O	O	-9-48i	-76+122i	$i(1+i)^{16}3^8$	$-2^4(2-i)^3(7+2i)^3/3^2$	I*5 , I*2
129*1+i	O	O	O	-9+3i	13-4i	$-(1+i)^8 3^7$	$-i(1+i)^{19}(2+i)^3/3$	III , I*1 
150*1	O	O	O	-9+48i	-76-122i	$-i(1+i)^{16}3^8$	$-2^4(2+i)^3(7-2i)^3/3^2$	I*5 , I*2

$$N=(1+i)^8(3)^2 = (144)$$

1*1	O	O	O	-3	O	$-(1+i)^{12}3^3$	$2^6 3^3$	I*O , III*
1*3	O	O	O	-27	O	$-(1+i)^{12}3^9$	$2^6 3^3$	I*O , III

$94\,{}^{*}i$	o	o	o	9	o	$(1+i)^{12}\,3^{6}$	$2^{6}3^{3}$	I*0 , I*0 
$116\,{}^{*}i$	o	o	o	99	$378i$	$i(1+i)^{18}\,3^{6}$	$2^{3}3^{3}11^{3}$	I*6 , I*0
$128\,{}^{*}i(1+i)$	o	o	o	$9+3i$	$4-13i$	$(1+i)^{8}\,3^{7}$	$-(1+i)^{19}(2-i)^{3}/3$	II , I*1 
$151\,{}^{*}1$	o	o	o	$9+48i$	$-122-76i$	$-i(1+i)^{16}\,3^{8}$	$-2^{4}(2-i)^{3}(7+2i)^{3}/3^{2}$	II* , I*2
$129\,{}^{*}i(1+i)$	o	o	o	$9-3i$	$-4-13i$	$(1+i)^{8}\,3^{7}$	$-i(1+i)^{19}(2+i)^{3}/3$	II , I*1
$150\,{}^{*}i$	o	o	o	$9-48i$	$-122+76i$	$i(1+i)^{16}\,3^{8}$	$-2^{4}(2+i)^{3}(7-2i)^{3}/3^{2}$	II* , I*2
$120\,{}^{*}1$	o	o	o	-39	-92	$-(1+i)^{12}\,3^{7}$	$2^{6}13^{3}/3$	I*0 , I*1 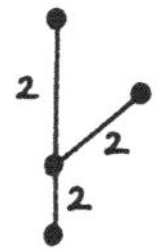
$138\,{}^{*}1$	$6i$	$9-15i$	$18-18i$	$162-54i$	$-3078i$	$i(1+i)^{18}\,3^{7}$	$2^{3}97^{3}/3$	I*6 , I*1
$143\,{}^{*}i$	o	o	o	21	$20i$	$(1+i)^{12}\,3^{8}$	$2^{6}7^{3}/3^{2}$	I*0 , I*2
$155\,{}^{*}1$	o	o	o	-69	$-146i$	$-i(1+i)^{18}\,3^{10}$	$2^{3}23^{3}/3^{4}$	I*6 , I*4
$127\,{}^{*}i(1+i)$	o	o	o	-6	$-7i$	$(1+i)^{8}\,3^{7}$	$2^{11}/3$	II , I*1 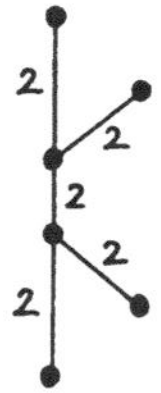
$140\,{}^{*}i$	$6i$	$9-6i$	$-18i$	513	$81+4212i$	$(1+i)^{20}\,3^{7}$	$2^{2}193^{3}/3$	I*8 , I*1
$149\,{}^{*}1$	o	o	o	39	$-70i$	$-(1+i)^{16}\,3^{8}$	$2^{4}13^{3}/3^{2}$	I*4 , I*2
$156\,{}^{*}i$	o	o	o	219	$1190i$	$(1+i)^{20}\,3^{10}$	$2^{2}73^{3}/3^{4}$	I*8 , I*4
$159\,{}^{*}i$	$6i$	$9+42i$	o	-729	o	$i(1+i)^{22}\,3^{14}$	$2\,47^{3}/3^{8}$	I*10, I*8
$152\,{}^{*}1$	$6i$	$9-24i$	108	$3267-324i$	$-2916-105462i$	$-i(1+i)^{22}\,3^{8}$	$2\,1153^{3}/3^{2}$	I*10, I*2
$9\,{}^{*}i(1+i)3$	o	o	o	o	$-27i$	$(1+i)^{8}\,3^{9}$	o	II , III* 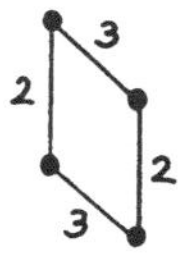
$9\,{}^{*}i(1+i)$	o	o	o	o	$-i$	$(1+i)^{8}\,3^{3}$	o	II , III
$21\,{}^{*}3$	o	o	o	135	$-594i$	$-(1+i)^{16}\,3^{9}$	$2^{4}3^{3}5^{3}$	I*4 , III*
$21\,{}^{*}1$	o	o	o	15	$-22i$	$-(1+i)^{16}\,3^{3}$	$2^{4}3^{3}5^{3}$	I*4 , III

$$N=(1+i)^9(3)^2=(144+144i)$$

									graph
148*1+i	o	o	o	30i	76-76i	$-i(1+i)^{20}3^8$	$2^5 5^3/3^2$	I*7 , I*2	•—•²
134*1+i	o	o	o	-15i	-13+13i	$-(1+i)^{10}3^7$	$2^7 5^3/3$	III , I*1	
148*i(1+i)	o	o	o	-30i	-76-76i	$i(1+i)^{20}3^8$	$2^5 5^3/3^2$	I*7 , I*2	•—•²
134*i(1+i)	o	o	o	15i	13+13i	$(1+i)^{10}3^7$	$2^7 5^3/3$	III , I*1	
158*1+i	o	o	o	-114i	860-860i	$-i(1+i)^{20}3^{12}$	$-2^5 19^3/3^6$	I*7 , I*6	•—•²
153*1+i	o	o	o	-159i	-545+545i	$-(1+i)^{10}3^9$	$2^7 53^3/3^3$	III , I*3	
158*i(1+i)	o	o	o	114i	-860-860i	$i(1+i)^{20}3^{12}$	$-2^5 19^3/3^6$	I*7 , I*6	•—•²
153*i(1+i)	o	o	o	159i	545+545i	$(1+i)^{10}3^9$	$2^7 53^3/3^3$	III , I*3	
11*(1+i)3	o	o	o	162i	540+540i	$-i(1+i)^{20}3^9$	$2^5 3^6$	I*7 , III*	•—•²
20*i(1+i)3	o	o	o	27i	27+27i	$(1+i)^{10}3^9$	$2^7 3^3$	III , III*	
11*i(1+i)3	o	o	o	-162i	540-540i	$i(1+i)^{20}3^9$	$2^5 3^6$	I*7 , III*	•—•²
20*(1+i)3	o	o	o	-27i	-27+27i	$-(1+i)^{10}3^9$	$2^7 3^3$	III , III*	
11*1+i	o	o	o	18i	20+20i	$-i(1+i)^{20}3^3$	$2^5 3^6$	I*7 , III	•—•²
20*i(1+i)	o	o	o	3i	1+i	$(1+i)^{10}3^3$	$2^7 3^3$	III , III	
11*i(1+i)	o	o	o	-18i	20-20i	$i(1+i)^{20}3^3$	$2^5 3^6$	I*7 , III	•—•²
20*1+i	o	o	o	-3i	-1+i	$-(1+i)^{10}3^3$	$2^7 3^3$	III , III	

105*1+i	o	o	o	42i	68+68i	$-i(1+i)^{20}3^6$	$2^5 7^3$	I*7 , I*O 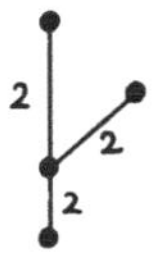
114*i	o	o	o	-45-48i	68+194i	$(1+i)^{17}3^6$	$(1+i)^7(2+3i)^3(6-i)^3$	I*4 , I*O
111*1+i	o	o	o	-3i	5+5i	$-(1+i)^{10}3^6$	2^7	III , I*O
115*i	o	o	o	45-48i	194+68i	$i(1+i)^{17}3^6$	$i(1+i)^7(2-3i)^3(6+i)^3$	II* , I*O
105*i(1+i)	o	o	o	-42i	68-68i	$i(1+i)^{20}3^6$	$2^5 7^3$	I*7 , I*O 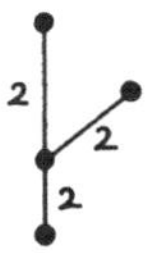
114*1	o	o	o	45+48i	-194+68i	$-(1+i)^{17}3^6$	$(1+i)^7(2+3i)^3(6-i)^3$	II* , I*O
111*i(1+i)	o	o	o	3i	5-5i	$(1+i)^{10}3^6$	2^7	III , I*O
115*1	o	o	o	-45+48i	-68+194i	$-i(1+i)^{17}3^6$	$i(1+i)^7(2-3i)^3(6+i)^3$	I*4 , I*O

$$N=(1+i)^{10}(3)^2=(288)$$

1*(1+i)3	o	o	o	-54i	o	$-(1+i)^{18}3^9$	$2^6 3^3$	I*4 , III*
1*i(1+i)3	o	o	o	54i	o	$(1+i)^{18}3^9$	$2^6 3^3$	I*4 , III*
1*1+i	o	o	o	-6i	o	$-(1+i)^{18}3^3$	$2^6 3^3$	I*4 , III
1*i(1+i)	o	o	o	6i	o	$(1+i)^{18}3^3$	$2^6 3^3$	I*4 , III
94*1+i	o	o	o	-18i	o	$-(1+i)^{18}3^6$	$2^6 3^3$	I*4 , I*O 
116*1+i	o	o	o	-198i	756-756i	$-i(1+i)^{24}3^6$	$2^3 3^3 11^3$	I*10, I*O
94*i(1+i)	o	o	o	18i	o	$(1+i)^{18}3^6$	$2^6 3^3$	I*4 , I*O 
116*i(1+i)	o	o	o	198i	-756-756i	$i(1+i)^{24}3^6$	$2^3 3^3 11^3$	I*10, I*O

128*1	o	o	o	6-18i	-34+18i	$-(1+i)^{14}3^7$	$-(4+i)^{19}(2-i)^3/3$	I*O , I*1	•–• 2
151*i(1+i)	o	o	o	96-18i	-92-396i	$i(1+i)^{22}3^8$	$-2^4(2-i)^3(7+2i)^3/3^2$	I*8 , I*2	
128*i	o	o	o	-6+18i	18+34i	$(1+i)^{14}3^7$	$-(4+i)^{19}(2-i)^3/3$	I*O , I*1	•–• 2
151*1+i	o	o	o	-96+18i	396-92i	$-i(1+i)^{22}3^8$	$-2^4(2-i)^3(7+2i)^3/3^2$	I*8 , I*2	
129*1	o	o	o	-6-18i	-18+34i	$-(1+i)^{14}3^7$	$-i(4+i)^{19}(2+i)^3/3$	I*O , I*1	•–• 2
150*1+i	o	o	o	-96-18i	396+92i	$-i(1+i)^{22}3^8$	$-2^4(2+i)^3(7-2i)^3/3^2$	I*8 , I*2	
129*i	o	o	o	6+18i	34+18i	$(1+i)^{14}3^7$	$-i(4+i)^{19}(2+i)^3/3$	I*O , I*1	•–• 2
150*i(1+i)	o	o	o	96+18i	92-396i	$i(1+i)^{22}3^8$	$-2^4(2+i)^3(7-2i)^3/3^2$	I*8 , I*2	
148*1	o	o	o	15	-38	$-i(1+i)^{14}3^8$	$2^5 5^3/3^2$	I*O , I*2	•–• 2
134*1	o	o	o	30	-52i	$-(1+i)^{16}3^7$	$2^7 5^3/3$	I*2 , I*1	
148*i	o	o	o	-15	38i	$i(1+i)^{14}3^8$	$2^5 5^3/3^2$	I*O , I*2	•–• 2
134*i	o	o	o	-30	-52	$(1+i)^{16}3^7$	$2^7 5^3/3$	I*2 , I*1	
158*1	o	o	o	-57	-430	$-i(1+i)^{14}3^{12}$	$-2^5 19^3/3^6$	I*O , I*6	•–• 2
153*1	o	-15i	o	243	-3645i	$-(1+i)^{16}3^9$	$2^7 53^3/3^3$	I*2 , I*3	
158*i	o	o	o	57	430i	$i(1+i)^{14}3^{12}$	$-2^5 19^3/3^6$	I*O , I*6	•–• 2
153*i	o	15	o	-243	-3645	$(1+i)^{16}3^9$	$2^7 53^3/3^3$	I*2 , I*3	
11*3	o	o	o	81	-270i	$-i(1+i)^{14}3^9$	$2^5 3^6$	I*O , III*	•–• 2
20*i3	o	o	o	-54	-108	$(1+i)^{16}3^9$	$2^7 3^3$	I*2 , III*	

11*i3	o	o	o	-81	-270	$i(1+i)^{14}3^9$	$2^5 3^6$	I*0 , III* 
20*3	o	o	o	54	-108i	$-(1+i)^{16}3^9$	$2^7 3^3$	I*2 , III*
11*1	o	o	o	9	-10i	$-i(1+i)^{14}3^3$	$2^5 3^6$	I*0 , III 
20*i	o	o	o	-6	-4	$(1+i)^{16}3^3$	$2^7 3^3$	I*2 , III
11*i	o	o	o	-9	-10	$i(1+i)^{14}3^3$	$2^5 3^6$	I*0 , III 
20*1	o	o	o	6	-4i	$-(1+i)^{16}3^3$	$2^7 3^3$	I*2 , III
120*1+i	o	o	o	-78i	184-184i	$-(1+i)^{18}3^7$	$2^6 13^3/3$	I*4 , I*1 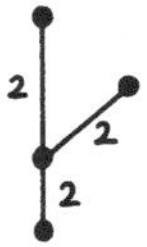
155*1+i	o	o	o	-138i	292+292i	$-i(1+i)^{24}3^{10}$	$2^3 23^3/3$	I*10, I*4
143*i(1+i)	o	o	o	42i	-40-40i	$(1+i)^{18}3^8$	$2^6 7^3/3^2$	I*4 , I*2
138*1+i	o	15-15i	72-18i	432i	5265+7128i	$i(1+i)^{24}3^7$	$2^3 97^3/3$	I*10, I*1
120*i(1+i)	o	o	o	78i	-184-184i	$(1+i)^{18}3^7$	$2^6 13^3/3$	I*4 , I*1 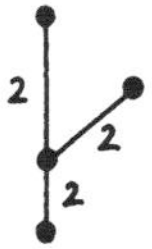
155*i(1+i)	o	o	o	138i	292-292i	$i(1+i)^{24}3^{10}$	$2^3 23^3/3^4$	I*10, I*4
143*1+i	o	o	o	-42i	40-40i	$-(1+i)^{18}3^8$	$2^6 7^3/3^2$	I*4 , I*2
138*i(1+i)	o	15-12i	18-18i	27-702i	1377-6804i	$-i(1+i)^{24}3^7$	$2^3 97^3/3$	I*10, I*1
105*1	o	o	o	21	-34i	$-i(1+i)^{14}3^6$	$2^5 7^3$	I*0 , I*0 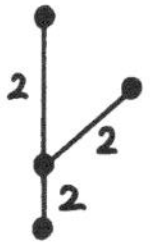
114*i(1+i)	o	o	o	96-90i	-524-252i	$(1+i)^{23}3^6$	$(1+i)^7(2+3i)^3(6-i)^3$	I*9 , I*0
111*1	o	o	o	6	-20	$-(1+i)^{16}3^6$	2^7	I*2 , I*0
115*i(1+i)	o	o	o	96+90i	-524+252i	$i(1+i)^{23}3^6$	$i(1+i)^7(2-3i)^3(6+i)^3$	I*9 , I*0

The four columns are headed by graphs (with edge weights 2; 2,2,2; 2,2,2,2; and a parallelogram 3,2,2,3). The data for each graph follow.

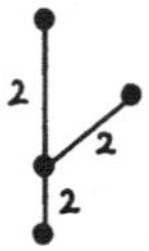

Graph 1

Fibres	Factorisation	Discriminant						Label
I*O , I*O	$2^5\,7^3$	$i(1+i)^{14}3^6$	-34	-21	0	0	0	105*i
I*9 , I*O	$(4+i)^3(2+3i)^3(6-i)^3$	$-(1+i)^{23}3^6$	252-524i	-96+90i	0	0	0	114*1+i
I*2 , I*O	2^7	$(1+i)^{16}3^6$	20i	-6	0	0	0	111*i
I*9 , I*O	$i(4+i)^3(2-3i)^3(6+i)^3$	$-i(1+i)^{23}3^6$	-252-524i	-96-90i	0	0	0	115*1+i

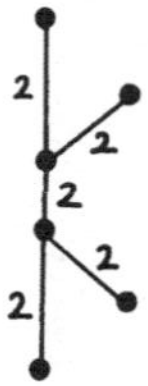

Graph 2

Fibres	Factorisation	Discriminant						Label
I*O , I*1	$2^{11}/3$	$-(1+i)^{14}3^7$	-14+14i	12i	0	0	0	127*1
I*12, I*1	$2^2\,193^3/3$	$-(1+i)^{26}3^7$	8667-19116i	135-1242i	0	21-6i	0	140*1+i
I*8 , I*2	$2^4\,13^3/3^2$	$(1+i)^{22}3^8$	140-140i	-78i	0	0	0	149*i(1+i)
I*12, I*4	$2^2\,73^3/3^4$	$-(1+i)^{26}3^{10}$	2380-2380i	-438i	0	0	0	156*1+i
I*14, I*8	$2\,47^3/3^8$	$-i(1+i)^{28}3^{14}$	0	1458i	0	42+42i	0	159*1+i
I*14, I*2	$2\,1153^3/3^2$	$i(1+i)^{28}3^8$	87723-150174i	-297-6858i	54+54i	-3-30i	0	152*i(1+i)

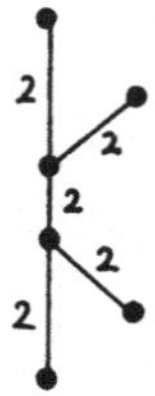

Graph 3

Fibres	Factorisation	Discriminant						Label
I*O , I*1	$2^{11}/3$	$(1+i)^{14}3^7$	14+14i	-12i	0	0	0	127*i
I*12, I*1	$2^2\,193^3/3$	$(1+i)^{26}3^7$	-8667-19116i	135+1242i	0	-21-6i	0	140*i(1+i)
I*8 , I*2	$2^4\,13^3/3^2$	$-(1+i)^{22}3^8$	140+140i	78i	0	0	0	149*1+i
I*12, I*4	$2^2\,73^3/3^4$	$(1+i)^{26}3^{10}$	-2380-2380i	438i	0	0	0	156*i(1+i)
1*14, I*8	$2\,47^3/3^8$	$i(1+i)^{28}3^{14}$	0	-1458i	0	-42+42i	0	159*i(1+i)
I*14, I*2	$2\,1153^3/3^2$	$-i(1+i)^{28}3^8$	217971+145800i	-189+6966i	54+162i	-3-24i	0	152*1+i

Graph 4

Fibres	Factorisation	Discriminant						Label
I*O , III*	0	$-(1+i)^{14}3^9$	-54+54i	0	0	0	0	9*3
I*O , III	0	$-(1+i)^{14}3^3$	-2+2i	0	0	0	0	9*1
I*8 , III*	$2^4\,3\,5^3$	$(1+i)^{22}3^9$	1188-1188i	-270i	0	0	0	21*i(1+i)3
I*8 , III	$2^3\,3\,5^3$	$(1+i)^{22}3^3$	44-44i	-30i	0	0	0	21*i(1+i)

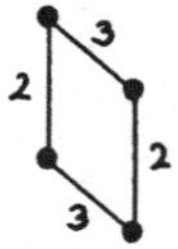

$9^{*}i3$	o	o	o	o	$54+54i$	$(1+i)^{14}3^{9}$	o	I*0 , III*
$9^{*}i$	o	o	o	o	$2+2i$	$(1+i)^{14}3^{3}$	o	I*0 , III
$21^{*}(1+i)3$	o	o	o	$270i$	$1188+1188i$	$-(1+i)^{22}3^{9}$	$2^{4}3^{3}5^{3}$	I*8 , III*
$21^{*}1+i$	o	o	o	$30i$	$44+44i$	$-(1+i)^{22}3^{3}$	$2^{4}3^{3}5^{3}$	I*8 , III

$$N=(1+i)^{12}(3)^{2}=(576)$$

$4^{*}1$	o	o	o	$3i$	o	$-i(1+i)^{12}3^{3}$	$2^{6}3^{3}$	II , III	•
$4^{*}i$	o	o	o	$-3i$	o	$i(1+i)^{12}3^{3}$	$2^{6}3^{3}$	II , III	•
$4^{*}3$	o	o	o	$27i$	o	$-i(1+i)^{12}3^{9}$	$2^{6}3^{3}$	II , III*	•
$4^{*}i3$	o	o	o	$-27i$	o	$i(1+i)^{12}3^{9}$	$2^{6}3^{3}$	II , III*	•
$4^{*}1+i$	o	o	o	-6	o	$-i(1+i)^{18}3^{3}$	$2^{6}3^{3}$	I*2 , III	•
$4^{*}i(1+i)$	o	o	o	6	o	$i(1+i)^{18}3^{3}$	$2^{6}3^{3}$	I*2 , III*	•
$4^{*}(1+i)3$	o	o	o	-54	o	$-i(1+i)^{18}3^{9}$	$2^{6}3^{3}$	I*2 , III*	•
$4\,i(1+i)3$	o	o	o	54	o	$i(1+i)^{18}3^{9}$	$2^{6}3^{3}$	I*2 , III*	•

95*1	O	O	O	9i	O	$-i(1+i)^{12}3^6$ $\quad 2^6 3^3$	II , I*O	•
95*i	O	O	O	-9i	O	$i(1+i)^{12}3^6$ $\quad 2^6 3^3$	II , I*O	•
95*1+i	O	O	O	-18	O	$-i(1+i)^{18}3^6$ $\quad 2^6 3^3$	I*2 , I*O	•
95*i(1+i)	O	O	O	18	O	$i(1+i)^{18}3^6$ $\quad 2^6 3^3$	I*2 , I*O	•
96*1	O	O	O	15i	-14-14i	$-i(1+i)^{12}3^6$ $\quad 20^3$	II , I*O	•—• 2
96*i	O	O	O	-15i	-14+14i	$i(1+i)^{12}3^6$ $\quad 20^3$	II , I*O	
96*1+i	O	O	O	-30	56	$-i(1+i)^{18}3^6$ $\quad 20^3$	I*2 , I*O	•—• 2
96*i(1+i)	O	O	O	30	-56i	$i(1+i)^{18}3^6$ $\quad 20^3$	I*2 , I*O	
154*1	O	O	O	-15i	-68+68i	$-i(1+i)^{12}3^{10}$ $\quad -2^6 5^3/3^4$	II , I*4	•—• 2
145*1	O	O	O	105i	-292-292i	$-i(1+i)^{12}3^8$ $\quad 2^6 5^3 7^3/3^2$	II , I*2	
154*i	O	O	O	15i	68+68i	$i(1+i)^{12}3^{10}$ $\quad -2^6 5^3/3^4$	II , I*4	•—• 2
145*i	O	O	O	-105i	-292+292i	$i(1+i)^{12}3^8$ $\quad 2^6 5^3 7^3/3^2$	II , I*2	
154*1+i	O	O	O	30	-272i	$-i(1+i)^{18}3^{10}$ $\quad -2^6 5^3/3^4$	I*2 , I*4	•—• 2
145*1+i	O	O	O	-210	1168	$-i(1+i)^{18}3^8$ $\quad 2^6 5^3 7^3/3^2$	I*2 , I*2	
154*i(1+i)	O	O	O	-30	-272	$i(1+i)^{18}3^{10}$ $\quad -2^6 5^3/3^4$	I*2 , I*4	•—• 2
145*i(1+i)	O	O	O	210	-1168i	$i(1+i)^{18}3^8$ $\quad 2^6 5^3 7^3/3^2$	I*2 , I*2	

123*1	0	0	0	33i	-50-50i	$-i(1+i)^{12}3^7$	$2^6 11^3/3$	II	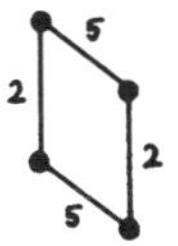 , I*1
157*1	0	-33+33i	0	2187i	-72171-72171i	$-i(1+i)^{12}3^{11}$	$2^6 971^3/3^5$	II	, I*5
144*1	0	0	0	3i	-22+22i	$-i(1+i)^{12}3^8$	$2^6/3^2$	II	, I*2
160*1	0	-66-66i	0	2187i	0	$-i(1+i)^{12}3^{16}$	$-2^6 239^3/3^{10}$	II	I*10
123*i	0	0	0	-33i	-50+50i	$i(1+i)^{12}3^7$	$2^6 11^3/3$	II	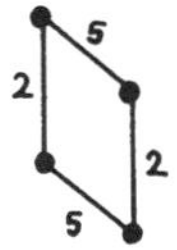 , I*1
157*i	0	-33-33i	0	-2187i	-72171+72171i	$i(1+i)^{12}3^{11}$	$2^6 971^3/3^5$	II	, I+5
144*i	0	0	0	-3i	22+22i	$i(1+i)^{12}3^8$	$2^6/3^2$	II	, I*2
160*i	0	66-66i	0	-2187i	0	$i(1+i)^{12}3^6$	$-2^6 239^3/3^{10}$	II	I*10
123*1+i	0	0	0	-66	200	$-i(1+i)^{18}3^7$	$2^6 11^3/3$	I*2	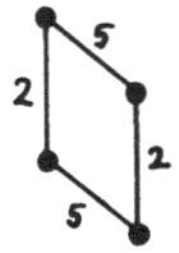 , I*1
157*1+i	0	-66	0	-4374	288684	$-i(1+i)^{18}3^{11}$	$2^6 971^3/3^5$	I*2	, I*5
144*1+i	0	0	0	-6	-88i	$-i(1+i)^{18}3^8$	$2^6/3^2$	I*2	, I*2
160*1+i	0	-132i	0	-4374	0	$-i(1+i)^{18}3^{16}$	$-2^6 239^3/3^{10}$	I*2	, I*10
123*i(1+i)	0	0	0	66	-200i	$i(1+i)^{18}3^7$	$2^6 11^3/3^5$	I*2	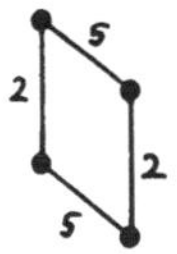 , I*1
157*i(1+i)	0	-66i	0	4374	-288684i	$i(1+i)^{18}3^{11}$	$2^6 971^3/3^5$	I*2	, I*5
144*i(1+i)	0	0	0	6	-88	$i(1+i)^{18}3^8$	$2^6/3^2$	I*2	, I*2
160*i(1+i)	0	132	0	4374	0	$-i(1+i)^{18}3^{16}$	$-2^6 239^3/3^{10}$	I*2	, I*10

$$N=(1+i)^{13}(3)^2=(576+576i)$$

103*1	0	0	0	3-9i	-18+2i	$-i(1+i)^{14}3^6$	$(1+i)^{13}(2-i)^3$	III	 , I*O
97*i	0	0	0	-12-9i	16+18i	$(1+i)^{13}3^6$	$-i(1+i)^{11}(2-i)^6$	II	, I*O

103^*i	O	O	O	$-3+9i$	$2+18i$	$i(1+i)^{14}3^6$	$(1+i)^{13}(2-i)^3$	III	, I*O	•–•$_2$
97^*1	O	O	O	$12+9i$	$-18+16i$	$-(1+i)^{13}3^6$	$-i(1+i)^{11}(2-i)^6$	II	, I*O	
103^*1+i	O	O	O	$18+6i$	$32-40i$	$-i(1+i)^{20}3^6$	$(1+i)^{13}(2-i)^3$	III*	, I*O	•–•$_2$
$97^*i(1+i)$	O	O	O	$18-24i$	$-68-4i$	$(1+i)^{19}3^6$	$-i(1+i)^{11}(2-i)^6$	I*2	, I*O	
$103^*i(1+i)$	O	O	O	$-18-6i$	$-40-32i$	$i(1+i)^{20}3^6$	$(1+i)^{13}(2-i)^3$	III*	, I*O	•–•$_2$
97^*1+i	O	O	O	$-18+24i$	$4-68i$	$-(1+i)^{19}3^6$	$-i(1+i)^{11}(2-i)^6$	I*2	, I*O	
146^*1	O	O	O	$-9-21i$	$-22+14i$	$-i(1+i)^{14}3^8$	$i(1+i)^{13}(5+2i)^3/3^2$	III	, I*2	•–•$_2$
125^*1	O	O	O	$36+39i$	$-140+50i$	$-(1+i)^{13}3^7$	$(1+i)^{11}(12+13i)^3/3$	II	, I*1	
146^*i	O	O	O	$9+21i$	$14+22i$	$i(1+i)^{14}3^8$	$i(1+i)^{13}(5+2i)^3/3^2$	III	, I*2	•–•$_2$
125^*i	O	O	O	$-36-39i$	$50+140i$	$(1+i)^{13}3^7$	$(1+i)^{11}(12+13i)^3/3$	II	, I*1	
146^*1+i	O	O	O	$42-18i$	$16-72i$	$-i(1+i)^{20}3^8$	$i(1+i)^{13}(5+2i)^3/3^2$	III*	, I*2	•–•$_2$
125^*1+i	O	O	O	$-78+72i$	$180-380i$	$-(1+i)^{19}3^7$	$(1+i)^{11}(12+13)^3/3$	I*2	, I*1	
$146^*i(1+i)$	O	O	O	$-42+18i$	$-72-16i$	$i(1+i)^{20}3^8$	$i(1+i)^{13}(5+2i)^3/3^2$	III	, I*2	•–•$_2$
$125^*i(1+i)$	O	O	O	$78-72i$	$-380-180i$	$(1+i)^{19}3^7$	$(1+i)^{11}(12+13i)^3/3$	I*2	, I*1	
147^*1	O	O	O	$-9+21i$	$-22-14i$	$-i(1+i)^{14}3^8$	$-(1+i)^{13}(5-2i)^3/3^2$	III	, I*2	•–•$_2$
126^*i	O	O	O	$36-39i$	$140+50i$	$i(1+i)^{13}3^7$	$-(1+i)^{11}(13+12i)^3/3$	II	, I*1	

147^*i	o	o	o	$9-21i$	$-14+22i$	$i(1+i)^{14}3^8$	$-(1+i)^{13}(5-2i)^3/3^2$	III , I*2	•—•²
126^*1	o	o	o	$-36+39i$	$-50+140i$	$-i(1+i)^{13}3^7$	$-(1+i)^{11}(13+12i)^3/3$	II , I*1	
147^*1+i	o	o	o	$-42-18i$	$72-16i$	$-i(1+i)^{20}3^8$	$-(1+i)^{13}(5-2i)^3/3^2$	III*, I*2	•—•²
$126^*i(1+i)$	o	o	o	$78+72i$	$-380+180i$	$i(1+i)^{19}3^7$	$-(1+i)^{11}(13+12i)^3/3$	I*2 , I*1	
$147^*i(1+i)$	o	o	o	$42+18i$	$-16-72i$	$i(1+i)^{20}3^8$	$-(1+i)^{13}(5-2i)^3/3^2$	III*, I*2	•—•²
126^*1+i	o	o	o	$-78-72i$	$-180-380i$	$-i(1+i)^{19}3^7$	$-(1+i)^{11}(13+12i)^3/3$	I*2 , I*1	
98^*1	o	o	o	$-12+9i$	$-16+18i$	$-i(1+i)^{13}3^6$	$-(1+i)^{11}(2+i)^6$	II , I*O	•—•²
104^*1	o	o	o	$3+9i$	$-18-2i$	$-i(1+i)^{14}3^6$	$-i(1+i)^{13}(2+i)^3$	III , I*O	
98^*i	o	o	o	$12-9i$	$18+16i$	$i(1+i)^{13}3^6$	$-(1+i)^{11}(2+i)^6$	II , I*O	•—•²
104^*i	o	o	o	$-3-9i$	$-2+18i$	$i(1+i)^{14}3^6$	$-i(1+i)^{13}(2+i)^3$	III , I*O	
98^*1+i	o	o	o	$-18-24i$	$-4-68i$	$-i(1+i)^{19}3^6$	$-(1+i)^{11}(2+i)^6$	I*2 , I*O	•—•²
104^*1+i	o	o	o	$-18+6i$	$40-32i$	$-i(1+i)^{20}3^6$	$-i(1+i)^{13}(2+i)^3$	III *, I*O	
$98^*i(1+i)$	o	o	o	$18+24i$	$-68+4i$	$i(1+i)^{19}3^6$	$-(1+i)^{11}(2+i)^6$	I*2 , I*O	•—•²
$104^*i(1+i)$	o	o	o	$18-6i$	$-32-40i$	$i(1+i)^{20}3^6$	$-i(1+i)^{13}(2+i)^3$	III*, I *O	

$$N=(1+i)^{14}(3)^2=(1152)$$

14*1	o	o	o	$-3-3i$	o	$-(1+i)^{15}3^3$	$2^6 3^3$	III , III	•
14*i	o	o	o	$3+3i$	o	$(1+i)^{15}3^3$	$2^6 3^3$	III , III	•
14*1+i	o	o	o	$6-6i$	o	$-(1+i)^{21}3^3$	$2^6 3^3$	III*, III	•
14*i(1+i)	o	o	o	$-6+6i$	o	$(1+i)^{21}3^3$	$2^6 3^3$	III*, III	•
14*3	o	o	o	$-27-27i$	o	$-(1+i)^{15}3^9$	$2^6 3^3$	III , III*	•
14*i3	o	o	o	$27+27i$	o	$(1+i)^{15}3^9$	$2^6 3^3$	III , III*	•
14*(1+i)3	o	o	o	$54-54i$	o	$-(1+i)^{21}3^9$	$2^6 3^3$	III*, III*	•
14*i(1+i)3	o	o	o	$-54+54i$	o	$(1+i)^{21}3^9$	$2^6 3^3$	III*, III*	•
17*1	o	o	o	$-3+3i$	o	$-i(1+i)^{15}3^3$	$2^6 3^3$	III , III	•
17*i	o	o	o	$3-3i$	o	$i(1+i)^{15}3^3$	$2^6 3^3$	III , III	•
17*1+i	o	o	o	$-6-6i$	o	$-i(1+i)^{21}3^3$	$2^6 3^3$	III*, III	•
17*i(1+i)	o	o	o	$6+6i$	o	$i(1+i)^{21}3^3$	$2^6 3^3$	III*, III	•

17^*3	o	o	o	$-27+27i$	o	$-i(1+i)^{15}3^9$	$2^6 3^3$	III , III*	●
17^*i3	o	o	o	$27-27i$	o	$i(1+i)^{15}3^9$	$2^6 3^3$	III , III*	●
$17^*(1+i)3$	o	o	o	$-54-54i$	o	$-i(1+i)^{21}3^9$	$2^6 3^3$	III*, III*	●
$17^*i(1+i)3$	o	o	o	$54+54i$	o	$i(1+i)^{21}3^9$	$2^6 3^3$	III*, III*	●
106^*1	o	o	o	$-9-9i$	o	$-(1+i)^{15}3^6$	12^3	III , I*O	●
106^*i	o	o	o	$9+9i$	o	$(1+i)^{15}3^6$	12^3	III , I*O	●
106^*1+i	o	o	o	$18-18i$	o	$-(1+i)^{21}3^6$	12^3	III*, I*O	●
$106^*i(1+i)$	o	o	o	$-18+18i$	o	$(1+i)^{21}3^6$	12^3	III*, I*O	●
108^*1	o	o	o	$-9+9i$	o	$-i(1+i)^{15}3^6$	12^3	III , I*O	●
108^*i	o	o	o	$9-9i$	o	$i(1+i)^{15}3^6$	12^3	III , I*O	●
108^*1+i	o	o	o	$-18-18i$	o	$-i(1+i)^{21}3^6$	12^3	III*, I*O	●
$108^*i(1+i)$	o	o	o	$18+18i$	o	$i(1+i)^{21}3^6$	12^3	III*, I*O	●
107^*1	o	o	o	$9+15i$	$-32+4i$	$-(1+i)^{15}3^6$	$-2^6(4+i)^3$	III , I*O	●
109^*1	o	o	o	$9-15i$	$-32-4i$	$-i(1+i)^{15}3^6$	$-2^6(4-i)^3$	III , I*O	●

107^*i	O	O	O	$-9-15i$	$4+32i$	$(1+i)^{15}3^6$	$-2^6(4+i)^3$	III , I*O 
109^*i	O	O	O	$-9+15i$	$-4+32i$	$i(1+i)^{15}3^6$	$-2^6(4-i)^3$	III , I*O
107^*1+i	O	O	O	$-30+18i$	$56-72i$	$-(1+i)^{21}3^6$	$-2^6(4+i)^3$	III*, I*O 
109^*1+i	O	O	O	$30+18i$	$72-56i$	$-i(1+i)^{21}3^6$	$-2^6(4-i)^3$	III*, I*O
$107^*i(1+i)$	O	O	O	$30-18i$	$-72-56i$	$(1+i)^{21}3^6$	$-2^6(4+i)^3$	III*, I*O 
$109^*i(1+i)$	O	O	O	$-30-18i$	$-56-72i$	$i(1+i)^{21}3^6$	$-2^6(4-i)^3$	III*, I*O

$$N=(3)^3=(27)$$

63^*3	O	O	$-i$	-270	1708	-3^{11}	$-2^{15}3\cdot5^3$	IO , II* 
2^*3	O	O	$-i$	O	7	-3^9	O	IO , IV*
2^*1	O	O	-1	O	O	-3^3	O	IO , II
63^*1	O	O	-1	-30	63	-3^5	$-2^{15}3\cdot5^3$	IO , IV

$$N=(1+i)(3)^3=(27+27i)$$

30^*1	-1	-1	$-1-i$	$-2+i$	$1-2i$	$-(1+i)3^3$	$-i3^3(2+i)^3(5+2i)^3/(4+i)$	I1 , II 
33^*3	-1	-1	$-1-i$	$-2-14i$	$8-14i$	$-(1+i)^33^9$	$i3^3(3-2i)^3(2+i)^3/(1+i)^3$	I3 , IV*
90^*3	$-i$	1	$-1-i$	$134-284i$	$2152-338i$	$-(1+i)^93^{11}$	$i3(6+5i)^3(8-7i)^3/(1+i)^9$	I9 , II*
30^*3	$-i$	1	-1	$-15+13i$	$12-27i$	$-(1+i)3^9$	$-i3^3(2+i)^3(5+2i)^3/(1+i)$	I1 , IV* 
33^*1	$-i$	1	-1	$-2i$	$-i$	$-(1+i)^33^3$	$i3^3(3-2i)^3(2+i)^3/(1+i)^3$	I3 , II
90^*1	-1	-1	$-i$	$15-32i$	$75-2i$	$-(1+i)^93^5$	$i3(6+5i)^3(8-7i)^3/(1+i)^9$	I9 , IV
31^*i3	-1	-1	$-i$	$-15-14i$	$-12-27i$	$i(1+i)3^9$	$-3^3(2-i)^3(5-2i)^3/(1+i)$	I1 , IV* 
34^*1	$-i$	1	-1	i	i	$-i(1+i)^33^3$	$3^3(2-i)^3(3+2i)^3/(1+i)^3$	I3 , II
91^*i	$-i$	1	-1	$15+31i$	$-75-2i$	$i(1+i)^93^5$	$3(2-i)^3(6-5i)^3(8+7i)^3/(1+i)^9$	I9 , IV
31^*i	$-i$	1	$-1-i$	$-1-2i$	$-1-2i$	$i(1+i)3^3$	$-3^3(2-i)^3(5-2i)^3/(1+i)$	I1 , II 
34^*3	-1	-1	$-1-i$	$-2+13i$	$8+13i$	$-i(1+i)^33^9$	$3^3(2-i)^3(3+2i)^3/(1+i)^3$	I3 , IV*
91^*i3	-1	-1	$-1-i$	$133+283i$	$-2152-338i$	$i(1+i)^93^{11}$	$3(2-i)^3(6-5i)^3(8+7i)^3/(1+i)^9$	I9 , II*
32^*i3	$-i$	1	$-i$	-28	53	$i(1+i)^23^9$	$-3^317^3/2$	I2 , IV* 
35^*1	$-i$	1	$-i$	2	1	$-i(1+i)^63^3$	$-3^37^3/2^3$	I6 , II
93^*i	-1	-1	-1	-14	29	$i(1+i)^{18}3^5$	$-3\cdot73^3/2^9$	I18 , IV

$32^{*}i$	-1	-1	O	-3	3	$i(1+i)^{2}3^{3}$	$-3^{3}17^{3}/2$	I2 ,II
$35^{*}3$	-1	-1	O	12	8	$-i(1+i)^{6}3^{9}$	$3^{3}7^{3}/2^{3}$	I6 ,IV*
$93^{*}i3$	$-i$	1	O	-123	667	$i(1+i)^{18}3^{11}$	$-3\cdot73^{3}/2^{9}$	I18 ,IV

$$N=(1+i)^{2}(3)^{3}=(54)$$

$22^{*}3$	O	O	$-1-i$	O	$-14i$	$-(1+i)^{4}3^{9}$	O	IV ,IV*
$22^{*}1$	O	O	$-1-i$	O	$-i$	$-(1+i)^{4}3^{3}$	O	IV ,II
$28^{*}i3$	$-1-i$	i	$-1-i$	$33+26i$	$-108+77i$	$i(1+i)^{8}3^{9}$	$i2^{2}3^{3}(5+4i)^{3}$	IV* ,IV*
$29^{*}1$	$-1-i$	i	$-1-i$	$3-4i$	$-2-i$	$-i(1+i)^{8}3^{3}$	$-i2^{2}3^{3}(5-4i)^{3}$	IV* ,II
$28^{*}i$	$-1-i$	i	O	$3+3i$	$-5+4i$	$i(1+i)^{8}3^{3}$	$i2^{2}3^{3}(5+4i)^{3}$	IV* ,II
$29^{*}3$	$-1-i$	i	O	$33-27i$	$-81-44i$	$-i(1+i)^{8}3^{9}$	$-i2^{2}3^{3}(5-4i)^{3}$	IV* ,IV*

$$N=(1+i)^{3}(3)^{3}=(54+54i)$$

$77^{*}1$	O	O	$-1-i$	3	$-3i$	$-(1+i)^{4}3^{5}$	$-2^{10}3$	III ,IV
$77^{*}3$	O	O	$-1-i$	27	$-68i$	$-(1+i)^{4}3^{11}$	$-2^{10}3$	III ,II*
$85^{*}1$	$-1-i$	i	O	O	$-4i$	$-i(1+i)^{10}3^{5}$	$-2\cdot3$	III*,IV
$85*3$	$-1-i$	i	$-1-i$	$6-i$	$-112i$	$-i(1+i)^{10}3^{11}$	$-2\cdot3$	III*

$$N=(1+i)^6 (3)^3 = (216)$$

3^*1	o	o	o	6	$-6i$	$-(1+i)^{12}3^3$	$-2^9 3^3$	I*2 ,II	●
3^*3	o	o	o	54	$-162i$	$-(1+i)^{12}3^9$	$-2^9 3^3$	I*2 ,IV*	●
59^*1	o	o	o	-6	$-2i$	$-(1+i)^{12}3^5$	$2^9 3$	I*2 ,IV	●
59^*3	o	o	o	-54	$-54i$	$-(1+i)^{12}3^{11}$	$2^9 3$	I*2 ,II*	●
27^*i	$-1-i$	i	$-1-i$	$-i$	$-i$	$i(1+i)^6 3^3$	$-2^3 3^3$	II ,II	●
27^*i3	$-1-i$	i	o	6	$-17i$	$i(1+i)^6 3^9$	$-2^3 3^3$	II ,IV*	●

$$N=(1+i)^7 (3)^3 = (216+216i)$$

83^*1	$-1-i$	i	o	$-3-3i$	$2-3i$	$-(1+i)^7 3^5$	$(1+i)^5 3(2+i)^6$	II ,IV	●
83^*3	$-1-i$	i	$-1-i$	$-21-28i$	$27-58i$	$-(1+i)^7 3^{11}$	$(1+i)^5 3(2+i)^6$	II ,II*	●
84^*1	$-1-i$	i	o	$-3+3i$	$-2-3i$	$-i(1+i)^7 3^5$	$-i(1+i)^5 3(2-i)^6$	II ,IV	●
84^*3	$-1-i$	i	$-1-i$	$-21+26i$	$-27-58i$	$-i(1+i)^7 3^{11}$	$-i(1+i)^5 3(2-i)^6$	II ,II*	●

$$N=(1+i)^8 (3)^3 =(432)$$

3^*i	o	o	o	-6	-6	$(1+i)^{12}3^3$	$-2^9 3^3$	I*O ,II	●
3^*i3	o	o	o	-54	-162	$(1+i)^{12}3^9$	$-2^9 3^3$	I*O ,IV*	●
59^*i	o	o	o	6	-2	$(1+i)^{12}3^5$	$2^9 3$	I*O ,IV	●
59^*i3	o	o	o	54	-54	$(1+i)^{12}3^{11}$	$2^9 3$	I*O ,II*	●
77^*i	o	o	o	12	2Oi	$(1+i)^{16}3^5$	$-2^{10}3$	II* ,IV	●
77^*i3	o	o	o	108	54Oi	$(1+i)^{16}3^{11}$	$-2^{10}3$	II* ,II*	●
27^*1	o	o	o	3	-6i	$-i(1+i)^{18}3^3$	$-2^3 3^3$	I*6 ,II	●
27^*3	o	o	o	27	-162i	$-i(1+i)^{18}3^9$	$-2^3 3^3$	I*6 ,IV*	●
83^*i	o	o	o	-9-12i	-4+14i	$(1+i)^{19}3^5$	$(1+i)^5 3(2+i)^6$	I*7 ,IV	●
83^*i3	o	o	o	-81-108i	-108+378i	$(1+i)^{19}3^{11}$	$(1+i)^5 3(2+i)^6$	I*7 ,II*	●
84^*i	o	o	o	-9-12i	4+14i	$i(1+i)^{19}3^5$	$-i(1+i)^5 3(2-i)^6$	I*7 ,IV	●
84^*i3	o	o	o	-81+108i	108+378i	$i(1+i)^{19}3^{11}$	$-i(1+i)^5 3(2-i)^6$	I*7 ,II*	●

85^*i	o	o	o	3	$34i$	$i(1+i)^{22}3^5$	$-2\cdot3$	I*10,IV	•
85^*i3	o	o	o	27	$918i$	$i(1+i)^{22}3^{11}$	$-2\cdot3$	I*10,II*	•
22^*i	o	o	o	o	$4i$	$(1+i)^{16}3^3$	o	II* ,II	 •
22^*i3	o	o	o	o	$108i$	$(1+i)^{16}3^9$	o	II* ,IV*	•—3
28^*1	o	o	o	$15+12i$	$-28+18i$	$-i(1+i)^{20}3^3$	$i2^2 3^3(5+4i)^3$	I*8 ,II	 •
29^*i3	o	o	o	$135-108i$	$756+486i$	$i(1+i)^{20}3^9$	$-i2^2 3^3(5-4i)^3$	I*8 ,IV*	•—3
28^*3	o	o	o	$135+108i$	$-756+486i$	$-i(1+i)^{20}3^9$	$i2^2 3^3(5+4i)^3$	I*8 ,IV*	 •
29^*i	o	o	o	$15-12i$	$28+18i$	$i(1+i)^{20}3^9$	$-i2^2 3^3(5-4i)^3$	I*8 ,II	•—3
30^*i	o	o	o	$27-24i$	$72+38i$	$(1+i)^{25}3^3$	$-i3^3(2+i)^3(5+2i)^3/(4+i)$	I*13,II	•
33^*i3	o	o	o	$27+216i$	$1080+486i$	$(1+i)^{27}3^9$	$i3^3(3-2i)^3(2+i)^3/(4+i)^3$	I*15,IV*	•—3
90^*i3	$6i$	18	$-378-216i$	$-2754+5670i$	$-13365+108378i$	$(1+i)^{33}3^{11}$	$i3(2+i)^3(6+5i)^3(8-7i)^3/(4+i)^9$	I*21,II*	•—3
30^*i3	o	o	o	$243-216i$	$1944+1026i$	$(1+i)^{25}3^9$	$-i3^3(2+i)^3(5+2i)^3/(1+i)$	I*13,IV*	 •
33^*i	o	o	o	$3-24i$	$40+18i$	$(1+i)^{27}3^3$	$i3^3(3-2i)^3(2+i)^3/(4+i)^3$	I*15,II	•—3
90^*i	$2i$	-59	-60	$963+564i$	$-3528-5058i$	$(1+i)^{33}3^5$	$i3(2+i)^3(6+5i)^3(8-7i)^3/(4+i)^9$	I*21,IV	•—3
31^*1	o	o	o	$27+24i$	$-72+38i$	$-i(1+i)^{25}3^3$	$-3^3(2-i)^3(5-2i)^3/(4+i)$	I*13,II	 •
34^*i3	o	o	o	$27-216i$	$-1080+486i$	$i(1+i)^{27}3^9$	$3^3(2-i)^3(3+2i)^3/(4+i)^9$	I*15,IV*	•—3
91^*3	$6i$	-189	$-162-378i$	$9801-4050i$	$-134622+404352i$	$-i(1+i)^{33}3^{11}$	$3(2-i)^3(6-5i)^3(8+7i)^3/(4+i)^9$	I*21,II*	•—3

31^*3	0	0	0	$243+216i$	$-1944+1026i$	$-i(1+i)^{25}3^9$	$-3^3(2-i)^3(5-2i)^3/(4+i)$	I*13,IV* 
34^*i	0	0	0	$3-24i$	$-40+18i$	$i(1+i)^{27}3^3$	$3^3(2-i)^3(3+2i)^3/(4+i)^3$	I*15,II
91^*1	$2i$	61	$66i$	$1029-504i$	$3717-5058i$	$-i(1+i)^{33}3^5$	$3(2-i)^3(6-5i)^3(8+7i)^3/(4+i)^9$	I*21,IV
32^*1	0	0	0	51	$-142i$	$-i(1+i)^{26}3^3$	$-3^3 17^3/2$	I*14,II 
35^*i3	0	0	0	-189	$702i$	$i(1+i)^{30}3^9$	$-3^3 7^3/2^3$	I*18,IV*
93^*3	0	0	0	1971	$-44658i$	$-i(1+i)^{42}3^{11}$	$-3\cdot73^3/2^9$	I*30,II*
32^*3	0	0	0	459	$-3834i$	$-i(1+i)^{26}3^9$	$-3^3 17^3/2$	I*14,IV* 
35^*i	0	0	0	-21	$26i$	$i(1+i)^{30}3^3$	$-3^3 7^3/2^3$	I*18,II
93^*1	0	0	0	219	$-1654i$	$-i(1+i)^{42}3^5$	$-3\cdot73^3/2^9$	I*30,IV
63^*i3	0	0	0	-1080	-13662	$(1+i)^{12}3^{11}$	$-2^{15}3\cdot5^3$	I*0 ,II* 
2^*i3	0	0	0	0	-54	$(1+i)^{12}3^9$	0	I*0 ,IV*
2^*i	0	0	0	0	-2	$(1+i)^{12}3^3$	0	I*0 ,II
63^*i	0	0	0	-120	-506	$(1+i)^{12}3^5$	$-2^{15}3\cdot5^3$	I*0 ,IV

$$N=(1+i)^9(3)^3=(432+432i)$$

69^*1+i	0	0	0	$-42i$	$76-76i$	$-i(1+i)^{20}3^5$	$-2^5 3\cdot7^3$	I*7 ,IV
$69^*i(1+i)$	0	0	0	$42i$	$-76-76i$	$i(1+i)^{20}3^5$	$-2^5 3\cdot7^3$	I*7 ,IV
$69^*(1+i)3$	0	0	0	$-378i$	$2052-2052i$	$-i(1+i)^{20}3^{11}$	$-2^5 3\cdot7^3$	I*7 ,II*

$69^* i(1+i)3$	o	o	o	$378i$	$-2052-2052i$	$i(1+i)^{20} 3^{11}$	$-2^5 3 \cdot 7^3$	I*7 ,II*	•
$12^* 1+i$	o	o	o	$6i$	$4-4i$	$-i(1+i)^{20} 3^3$	$2^5 3^3$	I*7 ,II	•
$12^* i(1+i)$	o	o	o	$-6i$	$-4-4i$	$i(1+i)^{20} 3^3$	$2^5 3^3$	I*7 ,II	•
$12^* (1+i)3$	o	o	o	$54i$	$108-108i$	$-i(1+i)^{20} 3^9$	$2^5 3^3$	I*7 ,IV*	•
$12^* i(1+i)3$	o	o	o	$-54i$	$-108-108i$	$i(1+i)^{20} 3^9$	$2^5 3^3$	I*7 ,IV*	•
$23^* 1$	o	o	o	$-6+6i$	$4-8i$	$-(1+i)^{16} 3^3$	$(1+i)^{17} 3^3$	III*,II	•
$23^* i$	o	o	o	$6-6i$	$-8-4i$	$(1+i)^{16} 3^3$	$(1+i)^{17} 3^3$	III*,II	•
$23^* 3$	o	o	o	$-54+54i$	$108-216i$	$-(1+i)^{16} 3^9$	$(1+i)^{17} 3^3$	III*,IV*	•
$23^* i3$	o	o	o	$54-54i$	$-216-108i$	$(1+i)^{16} 3^9$	$(1+i)^{17} 3^3$	III*,IV*	•
$24^* 1$	o	o	o	$-6-6i$	$-4-8i$	$-(1+i)^{16} 3^3$	$-i(1+i)^{17} 3^3$	III*,II	•
$24^* i$	o	o	o	$6+6i$	$-8+4i$	$(1+i)^{16} 3^3$	$-i(1+i)^{17} 3^3$	III*,II	•
$24^* 3$	o	o	o	$-54-54i$	$-108-216i$	$-(1+i)^{16} 3^9$	$-i(1+i)^{17} 3^3$	III*,IV*	•
$24^* i3$	o	o	o	$54+54i$	$-216+108i$	$(1+i)^{16} 3^9$	$-i(1+i)^{17} 3^3$	III*,IV*	•
$25^* 1$	o	o	o	3	$-4-2i$	$-(1+i)^{17} 3^3$	$(1+i)^7 3^3$	I*4 ,II	•

25^*i	o	o	o	-3	$-2+4i$	$(1+i)^{17}3^3$	$(1+i)^7 3^3$	II* ,II	●
25^*3	o	o	o	27	$-108-54i$	$-(1+i)^{17}3^9$	$(1+i)^7 3^3$	I*4 ,IV*	●
25^*i3	o	o	o	-27	$-54+108i$	$(1+i)^{17}3^9$	$(1+i)^7 3^3$	II* ,IV*	●
26^*1	o	o	o	-3	$-2-4i$	$-i(1+i)^{17}3^3$	$i(1+i)^7 3^3$	II* ,II	●
26^*i	o	o	o	3	$-4+2i$	$i(1+i)^{17}3^3$	$i(1+i)^7 3^3$	I*4 ,II	●
26^*3	o	o	o	-27	$-54-108i$	$-i(1+i)^{17}3^9$	$i(1+i)^7 3^3$	II* ,IV*	●
26^*i3	o	o	o	27	$-108+54i$	$i(1+i)^{17}3^9$	$i(1+i)^7 3^3$	I*4 ,IV*	●

$$N=(1+i)^{10}(3)^3=(864)$$

3^*1+i	o	o	o	$12i$	$12+12i$	$-(1+i)^{18}3^3$	$-2^9 3^3$	II* ,II	●
$3^*i(1+i)$	o	o	o	$-12i$	$12-12i$	$(1+i)^{18}3^3$	$-2^9 3^3$	II* ,II	●
$3^*(1+i)3$	o	o	o	$108i$	$324+324i$	$-(1+i)^{18}3^9$	$-2^9 3^3$	II* ,IV*	●
$3^*i(1+i)3$	o	o	o	$-108i$	$324-324i$	$(1+i)^{18}3^9$	$-2^9 3^3$	II* ,IV*	●
59^*1+i	o	o	o	$-12i$	$4+4i$	$-(1+i)^{18}3^5$	$2^9 3$	II* ,IV	●

$59^*i(1+i)$	o	o	o	$12i$	$4-4i$	$(1+i)^{18}3^5$	2^93	II* ,IV	•
$59^*(1+i)3$	o	o	o	$-108i$	$108+108i$	$-(1+i)^{18}3^{11}$	2^93	II* ,II*	•
$59^*i(1+i)3$	o	o	o	$108i$	$108-108i$	$(1+i)^{18}3^{11}$	2^93	II* ,II*	•
69^*1	o	o	o	-21	-38	$-i(1+i)^{14}3^5$	$-2^53\cdot7^3$	I*O ,IV	•
69^*i	o	o	o	21	$38i$	$i(1+i)^{14}3^5$	$-2^53\cdot7^3$	I*O ,IV	•
69^*3	o	o	o	-189	-1026	$-i(1+i)^{14}3^{11}$	$-2^53\cdot7^3$	I*O ,II*	•
69^*i3	o	o	o	189	$1026i$	$i(1+i)^{14}3^{11}$	$-2^53\cdot7^3$	I*O ,II*	•
12^*1	o	o	o	3	-2	$-i(1+i)^{14}3^3$	2^53^3	I*O ,II	•
12^*i	o	o	o	-3	$2i$	$i(1+i)^{14}3^3$	2^53^3	I*O ,II	•
12^*3	o	o	o	27	-54	$-i(1+i)^{14}3^9$	2^53^3	I*O ,IV*	•
12^*i3	o	o	o	-27	$54i$	$i(1+i)^{14}3^9$	2^53^3	I*O ,IV*	•
23^*1+i	o	o	o	$3+3i$	$-3+i$	$-(1+i)^{10}3^3$	$(1+i)^{17}3^3$	II ,II	•
$23^*i(1+i)$	o	o	o	$-3-3i$	$1+3i$	$(1+i)^{10}3^3$	$(1+i)^{17}3^3$	II ,II	•

23*(1+i)3	o	o	o	27+27i	-81+27i	$-(1+i)^{10}3^9$	$(1+i)^{17}3^3$	II ,IV*	●
23*i(1+i)3	o	o	o	-27-27i	27+81i	$(1+i)^{10}3^9$	$(1+i)^{17}3^3$	II ,IV*	●
24*1+i	o	o	o	-3+3i	-1+3i	$-(1+i)^{10}3^3$	$-i(1+i)^{17}3^3$	II ,II	●
24*i(1+i)	o	o	o	3-3i	3+i	$(1+i)^{10}3^3$	$-i(1+i)^{17}3^3$	II ,II	●
24*(1+i)3	o	o	o	-27+27i	-27+81i	$-(1+i)^{10}3^9$	$-i(1+i)^{17}3^3$	II ,IV*	●
24*i(1+i)3	o	o	o	27-27i	81+27i	$(1+i)^{10}3^9$	$-i(1+i)^{17}3^3$	II ,IV*	●
77*1+i	o	o	o	6i	5+5i	$-(1+i)^{10}3^5$	$-2^{10}3$	II ,IV	●
77*i(1+i)	o	o	o	-6i	5-5i	$(1+i)^{10}3^5$	$-2^{10}3$	II ,IV	●
77*(1+i)3	o	o	o	54i	135+135i	$-(1+i)^{10}3^{11}$	$-2^{10}3$	II ,II*	●
77*i(1+i)3	o	o	o	-54i	135-135i	$(1+i)^{10}3^{11}$	$-2^{10}3$	II ,II*	●
25*1+i	o	o	o	6i	12-4i	$-(1+i)^{23}3^3$	$(1+i)^7 3^3$	I*9 ,II	●
25*i(1+i)	o	o	o	-6i	-4-12i	$(1+i)^{23}3^3$	$(1+i)^7 3^3$	I*9 ,II	●
25*(1+i)3	o	o	o	54i	324-108i	$-(1+i)^{23}3^9$	$(1+i)^7 3^3$	I*9 ,IV*	●

25^*i(1+i)3	o	o	o	$-54i$	$-108-324i$	$(1+i)^{23}3^9$	$(1+i)^7 3^3$	I*9 ,IV*	●
26^*1+i	o	o	o	$-6i$	$12+4i$	$-i(1+i)^{23}3^3$	$i(1+i)^7 3^3$	I*9 ,II	●
26^*i(1+i)	o	o	o	$6i$	$4-12i$	$i(1+i)^{23}3^3$	$i(1+i)^7 3^3$	I*9 ,II	●
26^*(1+i)3	o	o	o	$-54i$	$324+108i$	$-i(1+i)^{23}3^9$	$i(1+i)^7 3^3$	I*9 ,IV*	●
26^*i(1+i)3	o	o	o	$54i$	$108-324i$	$i(1+i)^{23}3^9$	$i(1+i)^7 3^3$	I*9 ,IV*	●
27^*1+i	o	o	o	$6i$	$12+12i$	$-i(1+i)^{24}3^3$	$-2^3 3^3$	I*10,II	●
27^*i(1+i)	o	o	o	$-6i$	$12-12i$	$i(1+i)^{24}3^3$	$-2^3 3^3$	I*10,II	●
27^*(1+i)3	o	o	o	$54i$	$324+324i$	$-i(1+i)^{24}3^9$	$-2^3 3^3$	I*10,IV*	●
27^*i(1+i)3	o	o	o	$-54i$	$324-324i$	$i(1+i)^{24}3^9$	$-2^3 3^3$	I*10,IV*	●
83^*1+i	o	o	o	$-24+18i$	$36-20i$	$-(1+i)^{25}3^5$	$(1+i)^5 3(2+i)^6$	I*11,IV	●
83^*i(1+i)	o	o	o	$24-18i$	$-20-36i$	$(1+i)^{25}3^5$	$(1+i)^5 3(2+i)^6$	I*11,IV	●
83^*(1+i)3	o	o	o	$-216+162i$	$972-540i$	$-(1+i)^{25}3^{11}$	$(1+i)^5 3(2+i)^3$	I*11,II*	●
83^*i(1+i)3	o	o	o	$216-162i$	$-540-972i$	$(1+i)^{25}3^{11}$	$(1+i)^5 3(2+i)^3$	I*11,II*	●
84^*1+i	o	o	o	$24+18i$	$20-36i$	$-i(1+i)^{25}3^5$	$-i(1+i)^5 3(2-i)^6$	I*11,IV	●

$84^{*}i(1+i)$	O	O	O	$-24-18i$	$-36-20i$	$i(1+i)^{25}3^{5}$	$-i(1+i)^{5}3(2-i)^{6}$	I*11 ,IV	●
$84^{*}(1+i)3$	O	O	O	$216+162i$	$540-972i$	$-i(1+i)^{25}3^{11}$	$-i(1+i)^{5}3(2-i)^{6}$	I*11 ,II*	●
$84^{*}i(1+i)3$	O	O	O	$-216-162i$	$-972-540i$	$i(1+i)^{25}3^{11}$	$-i(1+i)^{5}3(2-i)^{6}$	I*11,II*	●
$85^{*}1+i$	O	O	O	$-6i$	$68-68i$	$-i(1+i)^{28}3^{5}$	$-2\cdot3$	I*14,IV	●
$85^{*}i(1+i)$	O	O	O	$6i$	$-68-68i$	$i(1+i)^{28}3^{5}$	$-2\cdot3$	I*14,IV	●
$85^{*}(1+i)3$	O	O	O	$-54i$	$1836-1836i$	$-i(1+i)^{28}3^{11}$	$-2\cdot3$	I*14,II*	●
$85^{*}i(1+i)3$	O	O	O	$54i$	$-1836-1836i$	$i(1+i)^{28}3^{11}$	$-2\cdot3$	I*14,II*	●
$22^{*}1+i$	O	O	O	O	$1+i$	$-(1+i)^{10}3^{3}$	O	II ,II	●—● 3
$22^{*}(1+i)3$	O	O	O	O	$27+27i$	$-(1+i)^{10}3^{9}$	O	II ,IV*	
$22^{*}i(1+i)$	O	O	O	O	$1-i$	$(1+i)^{10}3^{3}$	O	II ,II	●—● 3
$22^{*}i(1+i)3$	O	O	O	O	$27-27i$	$(1+i)^{10}3^{9}$	O	II ,IV*	
$28^{*}(1+i)3$	O	O	O	$-216+270i$	$540-2484i$	$-i(1+i)^{26}3^{9}$	$i2^{2}3^{3}(5+4i)^{3}$	I*12,IV*	●—● 3
$29^{*}i(1+i)$	O	O	O	$24+30i$	$-92+20i$	$i(1+i)^{26}3^{3}$	$-i2^{2}3^{3}(5-4i)^{3}$	I*12,II	
$28^{*}i(1+i)3$	O	O	O	$216-270i$	$-2484-540i$	$i(1+i)^{26}3^{9}$	$i2^{2}3^{3}(5+4i)^{3}$	I*12,IV*	●—● 3
$29^{*}1+i$	O	O	O	$-24-30i$	$-20-92i$	$-i(1+i)^{26}3^{3}$	$-i2^{2}3^{3}(5-4i)^{3}$	I*12,II	

28*1+i	0	0	0	-24+30i	20-92i	$-i(1+i)^{26}3^3$	$i2^2 3^3(5+4i)^3$	I*12,II 
29*i(1+i)3	0	0	0	216+270i	-2484+540i	$i(1+i)^{26}3^9$	$-i2^2 3^3(5-4i)^3$	I*12,IV*
28*i(1+i)	0	0	0	24-30i	-92-20i	$i(1+i)^{26}3^3$	$i2^2 3^3(5+4i)^3$	I*12,II 
29*(1+i)3	0	0	0	-216-270i	-540-2484i	$-i(1+i)^{26}3^9$	$-i2^2 3^3(5-4i)^3$	I*12,IV*
30*1+i	0	0	0	-48-54i	-68-220i	$-(1+i)^{31}3^3$	$-i3^3(2+i)^3(5+2i)^3/(1+i)$	I*17,II 
33*(1+i)3	0	0	0	432-54i	-1188-3132i	$-(1+i)^{33}3^9$	$i3^3(3-2i)^3(2+i)^3/(1+i)^3$	I*19,IV*
90*(1+i)3	0	-72-90i	378-756i	8100+8586i	305451-561816i	$-(1+i)^{39}3^{11}$	$i3(2+i)^3(6+5i)^3(8-7i)^3/(1+i)^9$	I*25,II*
30*i(1+i)	0	0	0	48+54i	-220+68i	$(1+i)^{31}3^3$	$-i3^3(2+i)^3(5+2i)^3/(1+i)$	I*17,II 
33*i(1+i)3	0	0	0	-432+54i	-3132+1188i	$(1+i)^{33}3^9$	$i3^3(3-2i)^3(2+i)^3/(1+i)^3$	I*19,IV*
90*i(1+i)3	0	-99-72i	-378i	-7533+486i	-51516+56862i	$(1+i)^{39}3^{11}$	$i3(2+i)^3(6+5i)^3(8-7i)^3/(1+i)^9$	I*25,II*
30*(1+i)3	0	0	0	-432-486i	-1836-5940i	$-(1+i)^{31}3^9$	$-i3^3(2+i)^3(5+2i)^3/(1+i)$	I*17,IV* 
33*1+i	0	0	0	48-6i	-44-116i	$-(1+i)^{33}3^3$	$i3^3(3-2i)^3(2+i)^3/(1+i)^3$	I*19,II
90*1+i	0	-150-30i	-126i	8208+3474i	-142911-119088i	$-(1+i)^{39}3^5$	$i3(2+i)^3(6+5i)^3(8-7i)^3/(1+i)^9$	I*25,IV
30*i(1+i)3	0	0	0	432+486i	-5940+1836i	$(1+i)^{31}3^9$	$-i3^3(2+i)^3(5+2i)^3/(1+i)$	I*17,IV* 
33*i(1+i)	0	0	0	-48+6i	-116+44i	$(1+i)^{33}3^3$	$i3^3(3-2i)^3(2+i)^3/(1+i)^3$	I*19,II
90*i(1+i)	0	228-150i	126-258i	8820-23274i	-229968-719550i	$(1+i)^{39}3^5$	$i3(2+i)^3(6+5i)^3(8-7i)^3/(1+i)^9$	I*25,IV
31*1+i	0	0	0	-48+54i	68-220i	$-i(1+i)^{31}3^3$	$-3^3(2-i)^3(5-2i)^3/(1+i)$	I*17,II 
34*i(1+i)3	0	0	0	432+54i	1188-3132i	$i(1+i)^{33}3^9$	$3^3(2-i)^3(3+2i)^3/(1+i)^3$	I*19,IV*
91*(1+i)3	0	72-90i	-378i	8100-8586i	-162567-704700i	$-i(1+i)^{39}3^{11}$	$3(2-i)^3(6-5i)^3(8-7i)^3/(1+i)^9$	I*25,II*

$31^*i(1+i)$	0	0	0	$48-54i$	$-220-68i$	$i(1+i)^{31}3^3$	$-3^3(2-i)^3(5-2i)^3/(4+i)$	I*17,II
$34^*(1+i)3$	0	0	0	$-432-54i$	$3132+1188i$	$-i(1+i)^{33}3^9$	$3^3(2-i)^3(3+2i)^3/(1+i)^3$	I*19,IV*
$91^*i(1+i)3$	0	$-99-306i$	$-378i$	$-37017+24462i$	$1458972+1749600i$	$i(1+i)^{39}3^{11}$	$3(2-i)^3(6-5i)^3(8+7i)^3/(4+i)^9$	I*25,II*
$31^*(1+i)3$	0	0	0	$-432+486i$	$1836-5940i$	$-i(1+i)^{31}3^9$	$-3^3(2-i)^3(5-2i)^3/(4+i)$	I*17,IV*
$34^*i(1+i)$	0	0	0	$48+6i$	$44-116i$	$i(1+i)^{33}3^3$	$3^3(2-i)^3(3+2i)^3/(4+i)^3$	I*19,II
91^*1+i	0	$150-417i$	$126-258i$	$-49455-42174i$	$-2784744+1484253i$	$-i(1+i)^{39}3^5$	$3(2-i)^3(6-5i)^3(8+7i)^3/(4+i)^9$	I*25,IV
$31^*i(1+i)3$	0	0	0	$432-486i$	$-5940-1836i$	$i(1+i)^{31}3^9$	$-3^3(2-i)^3(5-2i)^3/(4+i)$	I*17,IV*
34^*1+i	0	0	0	$-48-6i$	$116+44i$	$-i(1+i)^{33}3^3$	$3^3(2-i)^3(3+2i)^3/(4+i)^3$	I*19,II
$91^*i(1+i)$	0	$-159-228i$	$126-258i$	$-9909+24642i$	$860319-124956i$	$i(1+i)^{39}3^5$	$3(2-i)^3(6-5i)^3(8+7i)^3/(4+i)^9$	I*25,IV
32^*1+i	0	0	0	$102i$	$284+284i$	$-i(1+i)^{32}3^3$	$-3^3\,17^3/2$	I*18,II
$35^*i(1+i)3$	0	0	0	$-378i$	$-1404-1404i$	$i(1+i)^{36}3^9$	$-3^3\,7^3/2^3$	I*22,IV*
$93^*(1+i)3$	0	0	0	$3942i$	$89316+89316i$	$-i(1+i)^{48}3^{11}$	$-3\;73^3/2^9$	I*34,II*
$32^*i(1+i)$	0	0	0	$-102i$	$284-284i$	$i(1+i)^{32}3^3$	$-3^3\,17^3/2$	I*18,II
$35^*(1+i)3$	0	0	0	$378i$	$1404-1404i$	$-i(1+i)^{36}3^9$	$-3^3\,7^3/2^3$	I*22,IV*
$93^*i(1+i)3$	0	0	0	$-3942i$	$89316-89316i$	$i(1+i)^{48}3^{11}$	$-3\;73^3/2^9$	I*34,II*
$32^*(1+i)3$	0	0	0	$918i$	$7668+7668i$	$-i(1+i)^{32}3^9$	$-3^3\,17^3/2$	I*18,IV*
$35^*i(1+i)$	0	0	0	$-42i$	$-52-52i$	$i(1+i)^{36}3^3$	$-3^3\,7^3/2^3$	I*22,II
93^*1+i	0	0	0	$438i$	$3308+3308i$	$-i(1+i)^{48}3^5$	$-3\;73^3/2^9$	I*34,IV
$32^*i(1+i)3$	0	0	0	$-918i$	$7668-7668i$	$i(1+i)^{32}3^9$	$-3^3\,17^3/2$	I*18,IV*
35^*1+i	0	0	0	$42i$	$52-52i$	$-i(1+i)^{36}3^3$	$-3^3\,7^3/2^3$	I*22,II
$93^*i(1+i)$	0	0	0	$-438i$	$3308-3308i$	$i(1+i)^{48}3^5$	$-3\;73^3/2^9$	I*34,IV

$63^{*}(1+i)3$	o	o	o	$2160i$	$27324+27324i$	$-(1+i)^{18}3^{11}$	$-2^{15}3\cdot5^{3}$	II* ,II*
$2^{*}(1+i)3$	o	o	o	o	$108+108i$	$-(1+i)^{18}3^{9}$	o	II* ,IV*
$2^{*}1+i$	o	o	o	o	$4+4i$	$-(1+i)^{18}3^{3}$	o	II* ,II
$63^{*}1+i$	o	o	o	$240i$	$1012+1012i$	$-(1+i)^{18}3^{5}$	$-2^{15}3\cdot5^{3}$	II* ,IV
$63^{*}i(1+i)3$	o	o	o	$-2160i$	$27324-27324i$	$(1+i)^{18}3^{11}$	$-2^{15}3\cdot5^{3}$	II* ,II*
$2^{*}i(1+i)3$	o	o	o	o	$108-108i$	$(1+i)^{18}3^{9}$	o	II* ,IV*
$2^{*}i(1+i)$	o	o	o	o	$4-4i$	$(1+i)^{18}3^{3}$	o	II* ,II
$63^{*}i(1+i)$	o	o	o	$-240i$	$1012-1012i$	$(1+i)^{18}3^{5}$	$-2^{15}3\cdot5^{3}$	II* ,IV

$$N=(1+i)^{11}(3)^{3}=(864+864i)$$

$16^{*}1$	o	o	o	$-3+6i$	$-6i$	$-(1+i)^{15}3^{3}$	$-i(1+i)^{9}3^{3}(2+i)^{3}$	I*O ,II
$16^{*}i$	o	o	o	$3-6i$	-6	$(1+i)^{15}3^{3}$	$-i(1+i)^{9}3^{3}(2+i)^{3}$	I*O ,II
$16^{*}3$	o	o	o	$-27+54i$	$-162i$	$-(1+i)^{15}3^{9}$	$-i(1+i)^{9}3^{3}(2+i)^{3}$	I*O ,IV*
$16^{*}i3$	o	o	o	$27-54i$	-162	$(1+i)^{15}3^{9}$	$-i(1+i)^{9}3^{3}(2+i)^{3}$	I*O ,IV*
$16^{*}1+i$	o	o	o	$-12-16i$	$12+12i$	$-(1+i)^{21}3^{3}$	$-i(1+i)^{9}3^{3}(2+i)^{3}$	I*6 ,II
$16^{*}i(1+i)$	o	o	o	$12+6i$	$12-12i$	$(1+i)^{21}3^{3}$	$-i(1+i)^{9}3^{3}(2+i)^{3}$	I*6 ,II

$16^*(1+i)3$	o	o	o	$-108-54i$	$324+324i$	$-(1+i)^{21}3^9$	$-i(1+i)^9 3^3(2+i)^3$	I*6 ,IV*	●
$16^*i(1+i)3$	o	o	o	$108+54i$	$324-324i$	$(1+i)^{21}3^9$	$-i(1+i)^9 3^3(2+i)^3$	I*6 ,IV*	●
19^*1	o	o	o	$-3-6i$	$-6i$	$-i(1+i)^{15}3^3$	$-i(1+i)^9 3^3(2-i)^3$	I*O ,II	●
19^*i	o	o	o	$3+6i$	-6	$i(1+i)^{15}3^3$	$-i(1+i)^9 3^3(2-i)^3$	I*O ,II	●
19^*3	o	o	o	$-27-54i$	$-162i$	$-i(1+i)^{15}3^9$	$-i(1+i)^9 3^3(2-i)^3$	I*O ,IV*	●
19^*i3	o	o	o	$27+54i$	-162	$i(1+i)^{15}3^9$	$-i(1+i)^9 3^3(2-i)^3$	I*O ,IV*	●
19^*1+i	o	o	o	$12-6i$	$12+12i$	$-i(1+i)^{21}3^3$	$-i(1+i)^9 3^3(2-i)^3$	I*6 ,II	●
$19^*i(1+i)$	o	o	o	$-12+6i$	$12-12i$	$i(1+i)^{21}3^3$	$-i(1+i)^9 3^3(2-i)^3$	I*6 ,II	●
$19^*(1+i)3$	o	o	o	$108-54i$	$324+324i$	$-i(1+i)^{21}3^9$	$-i(1+i)^9 3^3(2-i)^3$	I*6 ,IV*	●
$19^*i(1+i)3$	o	o	o	$-108+54i$	$324-324i$	$i(1+i)^{21}3^9$	$-i(1+i)^9 3^3(2-i)^3$	I*6 ,IV*	●

$$N=(1+i)^{12}(3)^3=(1728)$$

5^*1	o	o	o	$-3i$	$-2+2i$	$-i(1+i)^{12}3^3$	$-2^6 3^3$	II ,II	●
5^*i	o	o	o	$3i$	$2+2i$	$i(1+i)^{12}3^3$	$-2^6 3^3$	II ,II	●

5^*3	o	o	o	$-27i$	$-54+54i$	$-i(1+i)^{12}3^9$	-2^63^3	II ,IV*	●
5^*i3	o	o	o	$27i$	$54+54i$	$i(1+i)^{12}3^9$	-2^63^3	II ,IV*	●
5^*1+i	o	o	o	6	$-8i$	$-i(1+i)^{18}3^3$	-2^63^3	I*2 ,II	●
$5^*i(1+i)$	o	o	o	-6	-8	$i(1+i)^{18}3^3$	-2^63^3	I*2 ,II	●
$5^*(1+i)3$	o	o	o	54	$-216i$	$-i(1+i)^{18}3^9$	-2^63^3	I*2 ,IV*	●
$5^*i(1+i)3$	o	o	o	-54	-216	$i(1+i)^{18}3^9$	-2^63^3	I*2 ,IV*	●
6^*1	o	o	o	$-69i$	$-156+156i$	$-i(1+i)^{12}3^3$	$-2^63^323^3$	II ,II	●
6^*i	o	o	o	$69i$	$156+156i$	$i(1+i)^{12}3^3$	$-2^63^323^3$	II ,II	●
6^*3	o	o	o	$-621i$	$-4212+4212i$	$-i(1+i)^{12}3^9$	$-2^63^323^3$	II ,IV*	●
6^*i3	o	o	o	$621i$	$4212+4212i$	$i(1+i)^{12}3^9$	$-2^63^323^3$	II ,IV*	●
6^*1+i	o	o	o	138	$-624i$	$-i(1+i)^{18}3^3$	$-2^63^323^3$	I*2 ,II	●
$6^*i(1+i)$	o	o	o	-138	-624	$i(1+i)^{18}3^3$	$-2^63^323^3$	I*2 ,II	●
$6^*(1+i)3$	o	o	o	1242	$-16848i$	$-i(1+i)^{18}3^9$	$-2^63^323^3$	I*2 ,IV*	●
$6^*i(1+i)3$	o	o	o	-1242	-16848	$i(1+i)^{18}3^9$	$-2^63^323^3$	I*2 ,IV*	●

$64^{*}1$	o	o	o	$3i$	$-4+4i$	$-i(1+i)^{12}3^{5}$	$2^{6}3$	II ,IV	●
$64^{*}i$	o	o	o	$-3i$	$4+4i$	$i(1+i)^{12}3^{5}$	$2^{6}3$	II ,IV	●
$64^{*}3$	o	o	o	$27i$	$-108+108i$	$-i(1+i)^{12}3^{11}$	$2^{6}3$	II ,II*	●
$64^{*}i3$	o	o	o	$-27i$	$108+108i$	$i(1+i)^{12}3^{11}$	$2^{6}3$	II ,II*	●
$64^{*}1+i$	o	o	o	-6	$-16i$	$-i(1+i)^{18}3^{5}$	$2^{6}3$	I*2 ,IV	●
$64^{*}i(1+i)$	o	o	o	6	-16	$i(1+i)^{18}3^{5}$	$2^{6}3$	I*2 ,IV	●
$64^{*}(1+i)3$	o	o	o	-54	$-432i$	$-i(1+i)^{18}3^{11}$	$2^{6}3$	I*2 ,II*	●
$64^{*}i(1+i)3$	o	o	o	54	-432	$i(1+i)^{18}3^{11}$	$2^{6}3$	I*2 ,II*	●

$$N=(1+i)^{13}(3)^3=(1728+1728i)$$

7^*1	o	o	o	$3i$	$-2i$	$-(1+i)^{13}3^3$	$-i(1+i)^{11}3^3$	II ,II	●
7^*i	o	o	o	$-3i$	-2	$(1+i)^{13}3^3$	$-i(1+i)^{11}3^3$	II ,II	●
7^*3	o	o	o	$27i$	$-54i$	$-(1+i)^{13}3^9$	$-i(1+i)^{11}3^3$	II ,IV*	●
7^*i3	o	o	o	$-27i$	-54	$(1+i)^{13}3^9$	$-i(1+i)^{11}3^3$	II ,IV*	●
7^*1+i	o	o	o	-6	$4+4i$	$-(1+i)^{19}3^3$	$-i(1+i)^{11}3^3$	I*2 ,II	●
$7^*i(1+i)$	o	o	o	6	$4-4i$	$(1+i)^{19}3^3$	$-i(1+i)^{11}3^3$	I*2 ,II	●
$7^*(1+i)3$	o	o	o	-54	$108+108i$	$-(1+i)^{19}3^9$	$-i(1+i)^{11}3^3$	I*2 ,IV*	●
$7^*i(1+i)3$	o	o	o	54	$108-108i$	$(1+i)^{19}3^9$	$-i(1+i)^{11}3^3$	I*2 ,IV*	●
8^*1	o	o	o	$3i$	-2	$-i(1+i)^{13}3^3$	$-(1+i)^{11}3^3$	II ,II	●
8^*i	o	o	o	$-3i$	$2i$	$i(1+i)^{13}3^3$	$-(1+i)^{11}3^3$	II ,II	●
8^*3	o	o	o	$27i$	-54	$-i(1+i)^{13}3^9$	$-(1+i)^{11}3^3$	II ,IV*	●
8^*i3	o	o	o	$-27i$	$54i$	$i(1+i)^{13}3^9$	$-(1+i)^{11}3^3$	II ,IV*	●
8^*1+i	o	o	o	-6	$4-4i$	$-i(1+i)^{19}3^3$	$-(1+i)^{11}3^3$	I*2 ,II	●

$8^{*}i(1+i)$	o	o	o	6	$-4-4i$	$i(1+i)^{19}3_3^{3}$	$-(1+i)^{11}3_3^{3}$	I*2 ,II	●
$8^{*}(1+i)3$	o	o	o	-54	$108-108i$	$-i(1+i)^{19}3_3^{9}$	$-(1+i)^{11}3_3^{3}$	I*2 ,IV*	●
$8^{*}i(1+i)3$	o	o	o	54	$-108-108i$	$i(1+i)^{19}3_3^{9}$	$-(1+i)^{11}3_3^{3}$	I*2 ,IV*	●
$10^{*}1$	o	o	o	$-3+3i$	$-2+2i$	$-i(1+i)^{14}3_3^{3}$	$i(1+i)^{13}3_3^{3}$	III ,II	●
$10^{*}i$	o	o	o	$3-3i$	$2+2i$	$i(1+i)^{14}3_3^{3}$	$i(1+i)^{13}3_3^{3}$	III ,II	●
$10^{*}3$	o	o	o	$-27+27i$	$-54+54i$	$-i(1+i)^{14}3_3^{9}$	$i(1+i)^{13}3_3^{3}$	III ,IV*	●
$10^{*}i3$	o	o	o	$27-27i$	$54+54i$	$i(1+i)^{14}3_3^{9}$	$i(1+i)^{13}3_3^{3}$	III ,IV*	●
$10^{*}1+i$	o	o	o	$-6-6i$	$-8i$	$-i(1+i)^{20}3_3^{3}$	$i(1+i)^{13}3_3^{3}$	III*,II	●
$10^{*}i(1+i)$	o	o	o	$6+6i$	-8	$i(1+i)^{20}3_3^{3}$	$i(1+i)^{13}3_3^{3}$	III*,II	●
$10^{*}(1+i)3$	o	o	o	$-54-54i$	$-216i$	$-i(1+i)^{20}3_3^{9}$	$i(1+i)^{13}3_3^{3}$	III*,IV*	●
$10^{*}i(1+i)3$	o	o	o	$54+54i$	-216	$i(1+i)^{20}3_3^{9}$	$i(1+i)^{13}3_3^{3}$	III*,IV*	●
$13^{*}1$	o	o	o	$-3-3i$	$-2-2i$	$-i(1+i)^{14}3_3^{3}$	$-i(1+i)^{13}3_3^{3}$	III ,II	●
$13^{*}i$	o	o	o	$3+3i$	$-2+2i$	$i(1+i)^{14}3_3^{3}$	$-i(1+i)^{13}3_3^{3}$	III ,II	●
$13^{*}3$	o	o	o	$-27-27i$	$-54-54i$	$-i(1+i)^{14}3_3^{9}$	$-i(1+i)^{13}3_3^{3}$	III ,IV*	●

13^*i3	o	o	o	$27+27i$	$-54+54i$	$i(1+i)^{14}3^9$	$-i(1+i)^{13}3^3$	III ,IV*	●
13^*1+i	o	o	o	$6-6i$	8	$-i(1+i)^{20}3^3$	$-i(1+i)^{13}3^3$	III*,II	●
$13^*i(1+i)$	o	o	o	$-6+6i$	$-8i$	$i(1+i)^{20}3^3$	$-i(1+i)^{13}3^3$	III*,II	●
$13^*(1+i)3$	o	o	o	$54-54i$	216	$-i(1+i)^{20}3^9$	$-i(1+i)^{13}3^3$	III*,IV*	●
$13^*i(1+i)3$	o	o	o	$-54+54i$	$-216i$	$i(1+i)^{20}3^9$	$-i(1+i)^{13}3^3$	III*,IV*	●
70^*1	o	o	o	$-3+45i$	$-74-90i$	$-i(1+i)^{14}3^5$	$i2^6 3(1+i)(7+8i)^3$	III ,IV	●
70^*i	o	o	o	$3-45i$	$-90+74i$	$i(1+i)^{14}3^5$	$i2^6 3(1+i)(7+8i)^3$	III ,IV	●
70^*3	o	o	o	$-27+405i$	$-1998-2430i$	$-i(1+i)^{14}3^{11}$	$i2^6 3(1+i)(7+8i)^3$	III ,II*	●
70^*i3	o	o	o	$27-405i$	$-2430+1998i$	$i(1+i)^{14}3^{11}$	$i2^6 3(1+i)(7+8i)^3$	III ,II*	●
70^*1+i	o	o	o	$-90-6i$	$328+32i$	$-i(1+i)^{20}3^5$	$i2^6 3(1+i)(7+8i)^3$	III*,IV	●
$70^*i(1+i)$	o	o	o	$90+6i$	$32-328i$	$i(1+i)^{20}3^5$	$i2^6 3(1+i)(7+8i)^3$	III*,IV	●
$70^*(1+i)3$	o	o	o	$-810-54i$	$8856+864i$	$-i(1+i)^{20}3^{11}$	$i2^6 3(1+i)(7+8i)^3$	III*,II*	●
$70^*i(1+i)3$	o	o	o	$810+54i$	$864-8856i$	$i(1+i)^{20}3^{11}$	$i2^6 3(1+i)(7+8i)^3$	III*,II*	●
71^*1	o	o	o	$-3-45i$	$-74+90i$	$-i(1+i)^{14}3^5$	$-i2^6 3(1+i)(8+7i)^3$	III ,IV	●

71^*i	o	o	o	$3+45i$	$90+74i$	$i(1+i)^{14}3^5$	$-i2^63(1+i)(8+7i)^3$	III ,IV	●
71^*3	o	o	o	$-27-405i$	$-1998+2430i$	$-i(1+i)^{14}3^{11}$	$-i2^63(1+i)(8+7i)^3$	III ,II*	●
71^*i3	o	o	o	$27+405i$	$2430+1998i$	$i(1+i)^{14}3^{11}$	$-i2^63(1+i)(8+7i)^3$	III ,II*	●
71^*1+i	o	o	o	$90-6i$	$-32-328i$	$-i(1+i)^{20}3^5$	$-i2^63(1+i)(8+7i)^3$	III*,IV	●
$71^*i(1+i)$	o	o	o	$-90+6i$	$-328+32i$	$i(1+i)^{20}3^5$	$-i2^63(1+i)(8+7i)^3$	III*,IV	●
$71^*(1+i)3$	o	o	o	$810-54i$	$-864-8856i$	$-i(1+i)^{20}3^{11}$	$-i2^63(1+i)(8+7i)^3$	III*,II*	●
$71^*i(1+i)3$	o	o	o	$-810+54i$	$-8856+864i$	$i(1+i)^{20}3^{11}$	$-i2^63(1+i)(8+7i)^3$	III*,II*	●

$$\underline{N=(1+i)^{14}(3)^3=(3456)}$$

15^*1	o	o	o	$3-3i$	-4	$-(1+i)^{15}3^3$	$-i2^63^3$	III ,II	●
15^*i	o	o	o	$-3+3i$	$4i$	$(1+i)^{15}3^3$	$-i2^63^3$	III ,II	●
15^*3	o	o	o	$27-27i$	-108	$-(1+i)^{15}3^9$	$-i2^63^3$	III ,IV*	●
15^*i3	o	o	o	$-27+27i$	$108i$	$(1+i)^{15}3^9$	$-i2^63^3$	III ,IV*	●

$15^{*}1+i$	o	o	o	$6+6i$	$8-8i$	$-(1+i)^{21}3^3$	$-i2^6 3^3$	III*,II	●
$15^{*}i(1+i)$	o	o	o	$-6-6i$	$-8-8i$	$(1+i)^{21}3^3$	$-i2^6 3^3$	III*,II	●
$15^{*}(1+i)3$	o	o	o	$54+54i$	$216-216i$	$-(1+i)^{21}3^9$	$-i2^6 3^3$	III*,IV*	●
$15^{*}i(1+i)3$	o	o	o	$-54-54i$	$-216-216i$	$(1+i)^{21}3^9$	$-i2^6 3^3$	III*,IV*	●
$18^{*}1$	o	o	o	$3+3i$	-4	$-i(1+i)^{15}3^3$	$i2^6 3^3$	III ,II	●
$18^{*}i$	o	o	o	$-3-3i$	$4i$	$i(1+i)^{15}3^3$	$i2^6 3^3$	III ,II	●
$18^{*}3$	o	o	o	$27+27i$	-108	$-i(1+i)^{15}3^9$	$i2^6 3^3$	III ,IV*	●
$18^{*}i3$	o	o	o	$-27-27i$	$108i$	$i(1+i)^{15}3^9$	$i2^6 3^3$	III ,IV*	●
$18^{*}1+i$	o	o	o	$-6+6i$	$8-8i$	$-i(1+i)^{21}3^3$	$i2^6 3^3$	III*,II	●
$18^{*}i(1+i)$	o	o	o	$6-6i$	$-8-8i$	$i(1+i)^{21}3^3$	$i2^6 3^3$	III*,II	●
$18^{*}(1+i)3$	o	o	o	$-54+54i$	$216-216i$	$-i(1+i)^{21}3^9$	$i2^6 3^3$	III*,IV*	●
$18^{*}i(1+i)3$	o	o	o	$54-54i$	$-216-216i$	$i(1+i)^{21}3^9$	$i2^6 3^3$	III*,IV*	●
$73^{*}1$	o	o	o	$9-3i$	$-4-4i$	$-(1+i)^{15}3^5$	$-i2^6 3(2+i)^3$	III ,IV	●
$73^{*}i$	o	o	o	$-9+3i$	$-4+4i$	$(1+i)^{15}3^5$	$-i2^6 3(2+i)^3$	III ,IV	●

73^*3	o	o	o	81-27i	-108-108i	$-(1+i)^{15}3^{11}$	$-i2^6 3(1+i)^3$	III ,II*	•
73^*i3	o	o	o	-81+27i	-108+108i	$(1+i)^{15}3^{11}$	$-i2^6 3(2+i)^3$	III ,II*	•
73^*1+i	o	o	o	6+18i	16	$-(1+i)^{21}3^5$	$-i2^6 3(2+i)^3$	III*,IV	•
$73^*i(1+i)$	o	o	o	-6-18i	-16i	$(1+i)^{21}3^5$	$-i2^6 3(2+i)^3$	III*,IV	•
$73^*(1+i)3$	o	o	o	54+162i	432	$-(1+i)^{21}3^{11}$	$-i2^6 3(2+i)^3$	III*,II*	•
$73^*i(1+i)3$	o	o	o	-54-162i	-432i	$(1+i)^{21}3^{11}$	$-i2^6 3(2+i)^3$	III*,II*	•
74^*1	o	o	o	9+3i	-4+4i	$-i(1+i)^{15}3^5$	$i2^6 3(2-i)^3$	III ,IV	•
74^*i	o	o	o	-9-3i	4+4i	$i(1+i)^{15}3^5$	$i2^6 3(2-i)^3$	III ,IV	•
74^*3	o	o	o	81+27i	-108+108i	$-i(1+i)^{15}3^{11}$	$i2^6 3(2-i)^3$	III ,II*	•
74^*i3	o	o	o	-81-27i	108+108i	$i(1+i)^{15}3^{11}$	$i2^6 3(2-i)^3$	III ,II*	•
74^*1+i	o	o	o	-6+13i	-16i	$-i(1+i)^{21}3^5$	$i2^6 3(2-i)^3$	III*,IV	•
$74^*i(1+i)$	o	o	o	6-18i	-16	$i(1+i)^{21}3^5$	$i2^6 3(2-i)^3$	III*,IV	•
$74^*(1+i)3$	o	o	o	-54+162i	-432i	$-i(1+i)^{21}3^{11}$	$i2^6 3(2-i)^3$	III*,II*	•
$74^*i(1+i)3$	o	o	o	54-162i	-432	$i(1+i)^{21}3^{11}$	$i2^6 3(2-i)^3$	III*,II*	•

$$N=(1+i)(3)^4=(81+81i)$$

118^*i	-1	-1	0	-6	8	$(1+i)^4 3^6$	$-3^3 11^3/2^2$	$I4$,IV
57^*i3	$-i$	1	0	39	19	$(1+i)^{12} 3^{10}$	$3^2 23^3/2^6$	$I12$,$IV*$
118^*i3	$-i$	1	$-i$	-55	161	$(1+i)^4 3^{12}$	$-3^3 11^3/2^2$	$I4$,$II*$
57^*i	-1	-1	-1	4	-1	$(1+i)^{12} 3^4$	$3^2 23^3/2^6$	$I12$,II
119^*1	-1	-1	0	-1077	13877	$-i(1+i)^{14} 3^6$	$-3^9 5^3 383^3/2^7$	$I14$,IV
117^*i	$-i$	1	0	3	1	$i(1+i)^2 3^6$	$3^6 5^3/2$	$I2$,IV
58^*i3	-1	-1	0	-852	19664	$i(1+i)^{42} 3^{10}$	$-3^2 5^3 101^3/2^{21}$	$I42$,$IV*$
56^*3	$-i$	1	0	-42	100	$-i(1+i)^6 3^{10}$	$-3^2 5^6/2^3$	$I6$,$IV*$
119^*3	$-i$	1	$-i$	-9694	364985	$-i(1+i)^{14} 3^{12}$	$-3^9 5^3 383^3/2^7$	$I14$,$II*$
117^*i3	-1	-1	-1	25	1	$i(1+i)^2 3^{12}$	$3^6 5^3/2$	$I2$,$II*$
58^*i	$-i$	1	$-i$	-49	791	$i(1+i)^{42} 3^4$	$-3^2 5^3 101^3/2^{21}$	$I42$,II
56^*1	-1	-1	-1	-5	5	$-i(1+i)^6 3^4$	$-3^2 5^6/2^3$	$I6$,II

$$M=(1+i)^2(3)^4=(162)$$

$43^*i(1+i)$	0	0	0	-21	37	$(1+i)^8 3^4$	$2^8 3^2 7^3$	$IV*$,II
$99^*i(1+i)3$	0	0	0	-81	243	$(1+i)^8 3^{12}$	$2^8 3^3$	$IV*$,$II*$
$43^*i(1+i)3$	0	0	0	-189	999	$(1+i)^8 3^{10}$	$2^8 3^2 7^3$	$IV*$,$IV*$
$99^*i(1+i)$	0	0	0	-9	9	$(1+i)^8 3^6$	$2^8 3^3$	$IV*$,IV

$51{*}1$	$-1-i$	i	0	9	$-7i$	$-(1+i)^4 3^4$	$-2^4 3^2 13^3$	IV ,II	●—● 3
$110{*}3$	$-1-i$	i	0	-21	$-71i$	$-(1+i)^4 3^{12}$	$2^4 3^3$	IV ,II*	
$51{*}3$	$-1-i$	i	$-1-i$	$87-i$	$-274i$	$-(1+i)^4 3^{10}$	$-2^4 3^2 13^3$	IV ,IV*	●—● 3
$110{*}1$	$-1-i$	i	$-1-i$	$-3-i$	$-4i$	$-(1+i)^4 3^6$	$2^4 3^3$	IV ,IV	

$$N=(1+i)^3 (3)^4=(162+162i)$$

$42{*}i(1+i)$	0	0	0	-3	1	$(1+i)^8 3^4$	$2^8 3^2$	I*1 ,II	●
$42{*}i(1+i)3$	0	0	0	-27	27	$(1+i)^8 3^{10}$	$2^8 3^2$	I*1 ,IV*	●
$53{*}i$	$-1-i$	i	$-1-i$	$-i$	$-2i$	$(1+i)^8 3^4$	$-2^2 3^2$	I*1 ,II	●
$53{*}i3$	$-1-i$	i	0	6	$-44i$	$(1+i)^8 3^{10}$	$-2^2 3^2$	I*1 ,IV*	●
$54{*}1$	$-1-i$	i	$-1-i$	$-6-i$	$-4-4i$	$-(1+i)^{11} 3^4$	$(1+i)3^2 7^3$	II* ,II	●
$54{*}3$	$-1-i$	i	0	-48	$-108-44i$	$-(1+i)^{11} 3^{10}$	$(1+i)3^2 7^3$	II* ,IV*	●
$55{*}1$	$-1-i$	i	$-1-i$	$-6-i$	$4-4i$	$-i(1+i)^{11} 3^4$	$-i(1+i)3^2 7^3$	II* ,II	●
$55{*}3$	$-1-i$	i	0	-48	$108-44i$	$-i(1+i)^{11} 3^{10}$	$-i(1+i)3^2 7^3$	II* ,IV*	●

$$N=(1+i)^6(3)^4=(648)$$

36^*1	o	o	o	3	$-4i$	$-(1+i)^{12}3^4$	-2^63^2	I*2 ,II	●
36^*3	o	o	o	27	$-108i$	$-(1+i)^{12}3^{10}$	-2^63^2	I*2 ,IV*	●
52^*i	$-1-i$	i	o	o	$-i$	$i(1+i)^63^4$	-2^33^2	II ,II	●
52^*i3	$-1-i$	i	$-1-i$	$6-i$	$-31i$	$i(1+i)^63^{10}$	-2^33^2	II ,IV*	●

$$N=(1+i)^7(3)^4=(648+648i)$$

44^*1+i	o	o	o	$3i$	$-2-i$	$-(1+i)^83^4$	2^83^2	III ,II	●
$44^*(1+i)3$	o	o	o	$27i$	$-54-27i$	$-(1+i)^83^{10}$	2^83^2	III ,IV*	●
45^*1+i	o	o	o	$-3i$	$2-i$	$-(1+i)^83^4$	$i2^83^2$	III ,II	●
$45^*(1+i)3$	o	o	o	$-27i$	$54-27i$	$-(1+i)^83^{10}$	$i2^83^2$	III ,IV*	●
46^*i	o	o	o	$-9+6i$	$-8+10i$	$(1+i)^{14}3^4$	$2^53^2(2+3i)^3$	III*,II	●
46^*i3	o	o	o	$-81+54i$	$-216+270i$	$(1+i)^{14}3^{10}$	$2^53^2(2+3i)^3$	III*,IV*	●

47^*1	o	o	o	$-9-6i$	$-8-10i$	$-(1+i)^{14}3^4$	$-2^5 3^2(2-3i)^3$	III*,II	•
47^*3	o	o	o	$-81-54i$	$-216-270i$	$-(1+i)^{14}3^{10}$	$-2^5 3^2(2-3i)^3$	III*,IV*	•

$$\underline{N=(1+i)^8(3)^4=(1296)}$$

36^*i	o	o	o	-3	-4	$(1+i)^{12}3^4$	$-2^6 3^2$	I*0 ,II	•
36^*i3	o	o	o	-27	-108	$(1+i)^{12}3^{10}$	$-2^6 3^2$	I*0 ,IV*	•
42^*1+i	o	o	o	3	i	$-(1+i)^8 3^4$	$2^8 3^2$	II ,II	•
$42^*(1+i)3$	o	o	o	27	$27i$	$-(1+i)^8 3^{10}$	$2^8 3^2$	II ,IV*	•
$44^*i(1+i)$	o	o	o	$-3i$	$-1+2i$	$(1+i)^8 3^4$	$2^8 3^2$	II ,II	•
$44^*i(1+i)3$	o	o	o	$-27i$	$-27+54i$	$(1+i)^8 3^{10}$	$2^8 3^2$	II ,IV*	•
$45\ i(1+i)$	o	o	o	$3i$	$-1-2i$	$(1+i)^8 3^4$	$i2^8 3^2$	II ,II	•
$45^*i(1+i)3$	o	o	o	$27i$	$-27-54i$	$(1+i)^8 3^{10}$	$i2^8 3^2$	II ,IV*	•
46^*1	o	o	o	$9-6i$	$-10-8i$	$-(1+i)^{14}3^4$	$2^5 3^2(2+3i)^3$	I*2 ,II	•
46^*3	o	o	o	$81-54i$	$-270-216i$	$-(1+i)^{14}3^{10}$	$2^5 3^2(2+3i)^3$	I*2 ,IV*	•

47*i	o	o	o	9+6i	-10+8i	$(1+i)^{14}3^4$	$-2^5 3^2 (2-3i)^3$	I*2 ,II	•
47*i3	o	o	o	81+54i	-270+216i	$(1+i)^{14}3^{10}$	$-2^5 3^2 (2-3i)^3$	I*2 ,IV*	•
52*1	o	o	o	3	-10i	$-i(1+i)^{18}3^4$	$-2^3 3^2$	I*6 ,II	•
52*3	o	o	o	27	-270i	$-i(1+i)^{18}3^{10}$	$-2^3 3^2$	I*6 ,IV*	•
53*1	o	o	o	3	-14i	$-(1+i)^{20}3^4$	$-2^2 3^2$	I*8 ,II	•
53*3	o	o	o	27	-378i	$-(1+i)^{20}3^{10}$	$-2^2 3^2$	I*8 ,IV*	•
54*i	o	o	o	-21	32+6i	$(1+i)^{23}3^4$	$(1+i)3^2 7^3$	I*11,II	•
54*i3	o	o	o	-189	864+162i	$(1+i)^{23}3^{10}$	$(1+i)3^2 7^3$	I*11,IV*	•
55*i	o	o	o	-21	-32+6i	$i(1+i)^{23}3^4$	$-i(1+i)3^2 7^3$	I*11,II	•
55*i3	o	o	o	-189	-864+162i	$i(1+i)^{23}3^{10}$	$-i(1+i)3^2 7^3$	I*11,IV*	•
43*1+i	o	o	o	21	37i	$-(1+i)^8 3^4$	$2^8 3^2 7^3$	II ,II	•—• 3
99*(1+i)3	o	o	o	81	243i	$-(1+i)^8 3^{12}$	$2^8 3^3$	II ,II*	
43*(1+i)3	o	o	o	189	999i	$-(1+i)^{18}3^{10}$	$2^8 3^2 7^3$	II ,IV*	•—• 3
99*1+i	o	o	o	9	9i	$-(1+i)^8 3^6$	$2^8 3^3$	II ,IV	

51*i3	o	o	o	351	2538i	$(1+i)^{16}3^{10}$	$-2^4 3^2 13^3$	I*4 ,IV* 
110*i	o	o	o	-9	18i	$(1+i)^{16}3^6$	$2^4 3^3$	I*4 ,IV
51*i	o	o	o	39	94i	$(1+i)^{16}3^4$	$-2^4 3^2 13^3$	I*4 ,II 
110*i3	o	o	o	-81	486i	$(1+i)^{16}3^{12}$	$2^4 3^3$	I*4 ,II*
57*1	o	o	o	-69	-22i	$-(1+i)^{36}3^4$	$3^2 23^3/2^6$	I*24,II 
118*3	o	o	o	891	-11178i	$-(1+i)^{28}3^{12}$	$-3^3 11^3/2^2$	I*16,II*
57*3	o	o	o	-621	-594i	$-(1+i)^{36}3^{10}$	$3^2 23^3/2^6$	I*24,IV* 
118*1	o	o	o	99	-414i	$-(1+i)^{28}3^6$	$-3^3 11^3/2^2$	I*16,IV
119*i	2i	1+72i	-126-126i	15381+126i	1262772i	$i(1+i)^{38}3^6$	$-3^9 5^3 383^3/2^7$	I*26,IV 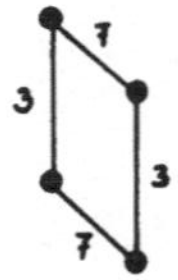
117*1	o	o	o	-45	-18i	$-i(1+i)^{26}3^6$	$3^6 5^3/2$	I*14,IV
58*3	o	o	o	13635	-1244862i	$-i(1+i)^{66}3^{10}$	$-3^2 5^3 101^3/2^{21}$	I*54,IV*
56*i3	o	o	o	675	7074i	$i(1+i)^{30}3^{10}$	$-3^2 5^6/2^3$	I*18,IV*
119*i3	6i	9-162i	1134-378i	145233-3402i	-285768 +15509718i	$i(1+i)^{38}3^{12}$	$-3^9 5^3 383^3/2^7$	I*26,II* 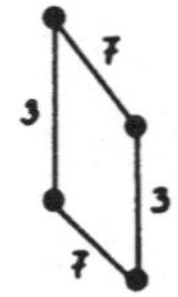
117*3	o	o	o	-405	-486i	$-i(1+i)^{26}3^{12}$	$3^6 5^3/2$	I*14,II*
58*1	o	o	o	1515	-46106i	$-i(1+i)^{66}3^4$	$-3^2 5^3 101^3/2^{21}$	I*54,II
56*i	o	o	o	75	262i	$i(1+i)^{30}3^4$	$-3^2 5^6/2^3$	I*18,II

$$N=(1+i)^9(3)^4=(1296+1296i)$$

50^*1+i	o	o	o	$-3i$	$1+i$	$-(1+i)^{10}3^4$	2^73^2	III ,II	●
$50^*i(1+i)$	o	o	o	$3i$	$1-i$	$(1+i)^{10}3^4$	2^73^2	III ,II	●
$50^*(1+i)3$	o	o	o	$-27i$	$27+27i$	$-(1+i)^{10}3^{10}$	2^73^2	III ,IV*	●
$50^*i(1+i)3$	o	o	o	$27i$	$27-27i$	$(1+i)^{10}3^{10}$	2^73^2	III ,IV*	●
100^*1+i	o	o	o	$-18i$	$36-36i$	$-i(1+i)^{20}3^6$	-2^53^3	I*7 ,IV	●
$100^*i(1+i)$	o	o	o	$18i$	$-36-36i$	$i(1+i)^{20}3^6$	-2^53^3	I*7 ,IV	●
$100^*(1+i)3$	o	o	o	$-162i$	$972-972i$	$-i(1+i)^{20}3^{12}$	-2^53^3	I*7 ,II*	●
$100^*i(1+i)3$	o	o	o	$162i$	$-972-972i$	$i(1+i)^{20}3^{12}$	-2^53^3	I*7 ,II*	●
112^*1+i	o	o	o	$9i$	$9+9i$	$-(1+i)^{10}3^6$	-2^73^3	III ,IV	●
$112^*i(1+i)$	o	o	o	$-9i$	$9-9i$	$(1+i)^{10}3^6$	-2^73^3	III ,IV	●
$112^*(1+i)3$	o	o	o	$81i$	$243+243i$	$-(1+i)^{10}3^{12}$	-2^73^3	III ,II*	●

$112^*i(1+i)3$	o	o	o	$-81i$	$243-243i$	$(1+i)^{10}3^{12}$	-2^73^3	III ,II*	●
113^*1+i	o	o	o	$549i$	$3501+3501i$	$-(1+i)^{10}3^6$	$-2^73^361^3$	III ,IV	●
$113^*i(1+i)$	o	o	o	$-549i$	$3501-3501i$	$(1+i)^{10}3^6$	$-2^73^361^3$	III ,IV	●
$113^*(1+i)3$	o	o	o	$4941i$	$94527+94527i$	$-(1+i)^{10}3^{12}$	$-2^73^361^3$	III ,II*	●
$113^*i(1+i)3$	o	o	o	$-4941i$	$94527-94527i$	$(1+i)^{10}3^{12}$	$-2^73^361^3$	III ,II*	●

$$N=(1+i)^{10}(3)^4=(2592)$$

36^*1+i	o	o	o	$6i$	$8+8i$	$-(1+i)^{18}3^4$	-2^63^2	I*4 ,II	●
$36^*i(1+i)$	o	o	o	$-6i$	$8-8i$	$(1+i)^{18}3^4$	-2^63^2	I*4 ,II	●
$36^*(1+i)3$	o	o	o	$54i$	$216+216i$	$-(1+i)^{18}3^{10}$	-2^63^2	I*4 ,IV*	●
$36^*i(1+i)3$	o	o	o	$-54i$	$216-216i$	$(1+i)^{18}3^{10}$	-2^63^2	I*4 ,IV*	●
42^*1	o	o	o	$6i$	$-2-2i$	$-(1+i)^{14}3^4$	2^83^2	I*O ,II	●
42^*i	o	o	o	$-6i$	$-2+2i$	$(1+i)^{14}3^4$	2^83^2	I*O ,II	●
42^*3	o	o	o	$54i$	$-54-54i$	$-(1+i)^{14}3^{10}$	2^83^2	I*O ,IV*	●

$42^{*}i3$	o	o	o	$-54i$	$-54+54i$	$(1+i)^{14}3^{10}$	$2^{8}3^{2}$	I*O ,IV*	●
$44^{*}1$	o	o	o	-6	$6-2i$	$-(1+i)^{14}3^{4}$	$2^{8}3^{2}$	I*O ,II	●
$44^{*}i$	o	o	o	6	$-2-6i$	$(1+i)^{14}3^{4}$	$2^{8}3^{2}$	I*O ,II	●
$44^{*}3$	o	o	o	-54	$162-54i$	$-(1+i)^{14}3^{10}$	$2^{8}3^{2}$	I*O ,IV*	●
$44^{*}i3$	o	o	o	54	$-54-162i$	$(1+i)^{14}3^{10}$	$2^{8}3^{2}$	I*O ,IV*	●
$45^{*}1$	o	o	o	6	$-2+6i$	$-(1+i)^{14}3^{4}$	$i2^{8}3^{2}$	I*O ,II	●
$45^{*}i$	o	o	o	-6	$6+2i$	$(1+i)^{14}3^{4}$	$i2^{8}3^{2}$	I*O ,II	●
$45^{*}3$	o	o	o	54	$-54+162i$	$-(1+i)^{14}3^{10}$	$i2^{8}3^{2}$	I*O ,IV*	●
$45^{*}i3$	o	o	o	-54	$162+54i$	$(1+i)^{14}3^{10}$	$i2^{8}3^{2}$	I*O ,IV*	●
$46^{*}1+i$	o	o	o	$12+18i$	$36-4i$	$-(1+i)^{20}3^{4}$	$2^{5}3^{2}(2+3i)^{3}$	I*6 ,II	●
$46^{*}i(1+i)$	o	o	o	$-12-18i$	$-4-36i$	$(1+i)^{20}3^{4}$	$2^{5}3^{2}(2+3i)^{3}$	I*6 ,II	●
$46^{*}(1+i)3$	o	o	o	$108+162i$	$972-108i$	$-(1+i)^{20}3^{10}$	$2^{5}3^{2}(2+3i)^{3}$	I*6 ,IV*	●
$46^{*}i(1+i)3$	o	o	o	$-108-162i$	$-108-972i$	$(1+i)^{20}3^{10}$	$2^{5}3^{2}(2+3i)^{3}$	I*6 ,IV*	●
$47^{*}1+i$	o	o	o	$12-18i$	$36+4i$	$-(1+i)^{20}3^{4}$	$-2^{5}3^{2}(2-3i)^{3}$	I*6 ,II	●

$47^*i(1+i)$	o	o	o	$-12+18i$	$4-36i$	$(1+i)^{20}3^4$	$-2^53^2(2-3i)^3$	I*6 ,II	●
$47^*(1+i)3$	o	o	o	$108-162i$	$972+108i$	$-(1+i)^{20}3^{10}$	$-2^53^2(2-3i)^3$	I*6 ,IV*	●
$47^*i(1+i)3$	o	o	o	$-108+162i$	$108-972i$	$(1+i)^{20}3^{10}$	$-2^53^2(2-3i)^3$	I*6 ,IV*	●
50^*1	o	o	o	6	-4	$-(1+i)^{16}3^4$	2^73^2	I*2 ,II	●
50^*i	o	o	o	-6	$4i$	$(1+i)^{16}3^4$	2^73^2	I*2 ,II	●
50^*3	o	o	o	54	-108	$-(1+i)^{16}3^{10}$	2^73^2	I*2 ,IV*	●
50^*i3	o	o	o	-54	$108i$	$(1+i)^{16}3^{10}$	2^73^2	I*2 ,IV*	●
52^*1+i	o	o	o	$6i$	$20+20i$	$-i(1+i)^{24}3^4$	-2^33^2	I*10,II	●
$52^*i(1+i)$	o	o	o	$-6i$	$20-20i$	$i(1+i)^{24}3^4$	-2^33^2	I*10,II	●
$52^*(1+i)3$	o	o	o	$54i$	$540+540i$	$-i(1+i)^{24}3^{10}$	-2^33^2	I*10,IV*	●
$52^*i(1+i)3$	o	o	o	$-54i$	$540-540i$	$i(1+i)^{24}3^{10}$	-2^33^2	I*10,IV*	●
53^*1+i	o	o	o	$6i$	$28+28i$	$-(1+i)^{26}3^4$	-2^23^2	I*12,II	●
$53^*i(1+i)$	o	o	o	$-6i$	$28-28i$	$(1+i)^{26}3^4$	-2^23^2	I*12,II	●
$53^*(1+i)3$	o	o	o	$54i$	$756+756i$	$-(1+i)^{26}3^{10}$	-2^23^2	I*12,IV*	●

$53^{*}\,i(1+i)3$	o	o	o	$-54i$	$756-756i$	$(1+i)^{26}3^{10}$	$-2^{2}3^{2}$	I*12,IV*	•
$54^{*}\,1+i$	o	o	o	$42i$	$-52-76i$	$-(1+i)^{29}3^{4}$	$(1+i)3^{2}7^{3}$	I*15,II	•
$54^{*}\,i(1+i)$	o	o	o	$-42i$	$-76+52i$	$(1+i)^{29}3^{4}$	$(1+i)^{29}3^{4}$	I*15,II	•
$54^{*}\,(1+i)3$	o	o	o	$378i$	$-1404-2052i$	$-(1+i)^{29}3^{10}$	$(1+i)3^{2}7^{3}$	I*15,IV*	•
$54^{*}\,i(1+i)3$	o	o	o	$-378i$	$-2052+1404i$	$(1+i)^{29}3^{10}$	$(1+i)3^{2}7^{3}$	I*15,IV*	•
$55^{*}\,1+i$	o	o	o	$42i$	$76+52i$	$-i(1+i)^{29}3^{4}$	$-i(1+i)3^{2}7^{3}$	I*15,II	•
$55^{*}\,i(1+i)$	o	o	o	$-42i$	$52-76i$	$i(1+i)^{29}3^{4}$	$-i(1+i)3^{2}7^{3}$	I*15,II	•
$55^{*}\,(1+i)3$	o	o	o	$378i$	$2052+1404i$	$-i(1+i)^{29}3^{10}$	$-i(1+i)3^{2}7^{3}$	I*15,IV*	•
$55^{*}\,i(1+i)3$	o	o	o	$-378i$	$1404-2052i$	$i(1+i)^{29}3^{10}$	$-i(1+i)3^{2}7^{3}$	I*15,IV*	•
$100^{*}\,1$	o	o	o	-9	-18	$-i(1+i)^{14}3^{6}$	$-2^{5}3^{3}$	I*0 ,IV	•
$100^{*}\,i$	o	o	o	9	$18i$	$i(1+i)^{14}3^{6}$	$-2^{5}3^{3}$	I*0 ,IV	•
$100^{*}\,3$	o	o	o	-81	-486	$-i(1+i)^{14}3^{12}$	$-2^{5}3^{3}$	I*0 ,II*	•
$100^{*}\,i3$	o	o	o	81	$486i$	$i(1+i)^{14}3^{12}$	$-2^{5}3^{3}$	I*0 ,II*	•
$112^{*}\,1$	o	o	o	-18	-36	$-(1+i)^{16}3^{6}$	$-2^{7}3^{3}$	I*2 ,IV	•

112^*i	o	o	o	18	36i	$(1+i)^{16}{}_3{}^6$	$-2^7 3^3$	I*2 ,IV	●
112^*3	o	o	o	-162	-972	$-(1+i)^{16}{}_3{}^{12}$	$-2^7 3^3$	I*2 ,II*	●
112^*i3	o	o	o	162	972i	$(1+i)^{16}{}_3{}^{12}$	$-2^7 3^3$	I*2 ,II*	●
113^*1	o	o	o	-1098	-14004	$-(1+i)^{16}{}_3{}^6$	$-2^7 3^3 61^3$	I*2 ,IV	●
113^*i	o	o	o	1098	14004i	$(1+i)^{16}{}_3{}^6$	$-2^7 3^3 61^3$	I*2 ,IV	●
113^*3	o	o	o	-9882	-378108	$-(1+i)^{16}{}_3{}^{12}$	$-2^7 3^3 61^3$	I*2 ,II*	●
113^*i3	o	o	o	9882	378108i	$(1+i)^{16}{}_3{}^{12}$	$-2^7 3^3 61^3$	I*2 ,II*	●
43^*1	o	o	o	42i	-74-74i	$-(1+i)^{14}{}_3{}^4$	$2^8 3^2 7^3$	I*O ,II	●—●3
99^*3	o	o	o	162i	-486-486i	$-(1+i)^{14}{}_3{}^{12}$	$2^8 3^3$	I*O ,II*	
43^*i	o	o	o	-42i	-74+74i	$(1+i)^{14}{}_3{}^4$	$2^8 3^2 7^3$	I*O ,II	●—●3
99^*i3	o	o	o	-162i	-486+486i	$(1+i)^{14}{}_3{}^{12}$	$2^8 3^3$	I*O ,II*	
43^*3	o	o	o	378i	-1998-1998i	$-(1+i)^{14}{}_3{}^{10}$	$2^8 3^2 7^3$	I*O ,IV*	●—●3
99^*1	o	o	o	18i	-18-18i	$-(1+i)^{14}{}_3{}^6$	$2^8 3^3$	I*O ,IV	
43^*i3	o	o	o	-378i	-1998+1998i	$(1+i)^{14}{}_3{}^{10}$	$2^8 3^2 7^3$	I*O ,IV*	●—●3
99^*i	o	o	o	-18i	-18+18i	$(1+i)^{14}{}_3{}^6$	$2^8 3^3$	I*O ,IV	

Each of the eight configurations carries the same graph ●—● 3, and each entry is marked below with ○ symbols.

Parameter	Fiber types	Discriminant factor	$(1+i)$ factor	Value	i-value
$51^{*}1{+}i$	$I_8^*,\,II$	$-2^4\,3^2\,13^3$	$-(1+i)22_3^{4}$	$188-188i$	$-78i$
$100^{*}(1{+}i)^{3}$	$I_8^*,\,II^*$	$2^4\,3^3$	$-(1+i)22_3^{12}$	$972-972i$	$162i$
$51^{*}i(1{+}i)$	$I_8^*,\,II$	$-2^4\,3^2\,13^3$	$(1+i)22_3^{4}$	$-188-188i$	$78i$
$110^{*}i(1{+}i)^{3}$	$I_8^*,\,II^*$	$2^4\,3^3$	$(1+i)22_3^{12}$	$-972-972i$	$-162i$
$51^{*}(1{+}i)^{3}$	$I_8^*,\,IV^*$	$-2^4\,3^2\,13^3$	$-(1+i)22_3^{10}$	$5076-5076i$	$-702i$
$110^{*}1{+}i$	$I_8^*,\,IV$	$2^4\,3^3$	$-(1+i)22_3^{6}$	$36-36i$	$18i$
$51^{*}i(1{+}i)^{3}$	$I_8^*,\,IV^*$	$-2^4\,3^2\,13^3$	$(1+i)22_3^{10}$	$-5076-5076i$	$702i$
$110^{*}i(1{+}i)$	$I_8^*,\,IV$	$2^4\,3^3$	$(1+i)22_3^{6}$	$-36-36i$	$-18i$
$57^{*}1{+}i$	$I_{28}^*,\,II$	$3^2\,23^3/2^6$	$-(1+i)42_3^{4}$	$44+44i$	$-138i$
$118^{*}(1{+}i)^{3}$	$I_{20}^*,\,II^*$	$-3^3\,11^3/2^2$	$-(1+i)34_3^{12}$	$22356+22356i$	$1782i$
$57^{*}i(1{+}i)$	$I_{28}^*,\,II$	$3^2\,23^3/2^6$	$(1+i)42_3^{4}$	$44-44i$	$138i$
$118^{*}i(1{+}i)^{3}$	$I_{20}^*,\,II^*$	$-3^3\,11^3/2^2$	$(1+i)34_3^{12}$	$22356-22356i$	$-1782i$
$57^{*}(1{+}i)^{3}$	$I_{28}^*,\,IV^*$	$3^2\,23^3/2^6$	$-(1+i)42_3^{10}$	$1188+1188i$	$-1242i$
$118^{*}1{+}i$	$I_{20}^*,\,IV$	$-3^3\,11^3/2^2$	$-(1+i)34_3^{6}$	$828+828i$	$198i$
$57^{*}i(1{+}i)^{3}$	$I_{28}^*,\,IV^*$	$3^2\,23^3/2^6$	$(1+i)42_3^{10}$	$1188-1188i$	$1242i$
$118^{*}i(1{+}i)$	$I_{20}^*,\,IV$	$-3^3\,11^3/2^2$	$(1+i)34_3^{6}$	$828-828i$	$-198i$

119*1+i	o	459+72i	258+768i	68499-12438i	6017076-5443146i	$-i(1+i)^{44}3^6$	$-3^9 5^3 383^3/2^7$	I*30,IV
117*i(1+i)	o	o	o	90i	36-36i	$i(1+i)^{32}3^6$	$3^6 5^3/2$	I*18,IV
58*i(1+i)3	o	o	o	-27270i	276636-276636i	$i(1+i)^{72}3^{10}$	$-3^2 5^3 101^3/2^{21}$	I*58,IV*
56*(1+i)3	o	o	o	-1350i	14148-14148i	$-i(1+i)^{36}3^{10}$	$-3^2 5^6/2^3$	I*22,IV*
119*i(1+i)	o	306+72i	258+768i	29484+49158i	-1553301+2410344i	$i(1+i)^{44}3^6$	$-3^9 5^3 383^3/2^7$	I*30,IV
117*1+i	o	o	o	-90i	36+36i	$-i(1+i)^{32}3^6$	$3^6 5^3/2$	I*18,IV
58*(1+i)3	o	o	o	27270i	276636+276636i	$-i(1+i)^{72}3^{10}$	$-3^2 5^3 101^3/2^{21}$	I*58,IV*
56*i(1+i)3	o	o	o	1350i	-14148-14148i	$i(1+i)^{36}3^{10}$	$-3^2 5^6/2^3$	I*22,IV*
119*(1+i)3	o	621+216i	756+756i	112995-220806i	75015315-102649518i	$-i(1+i)^{44}3^{12}$	$-3^9 5^3 383^3/2^7$	I*30,II*
117*i(1+i)3	o	o	o	810i	972-972i	$i(1+i)^{32}3^{12}$	$3^6 5^3/2$	I*18,II*
58*i(1+i)	o	o	o	-3030i	92212-92212i	$i(1+i)^{72}3^4$	$-3^2 5^3 101^3/2^{21}$	I*58,II
56*1+i	o	o	o	-150i	524-524i	$-i(1+i)^{36}3^4$	$-3^2 5^6/2^3$	I*22,II
119*i(1+i)3	o	972+225i	256+256i	269568+449280i	-46006272+6986376i	$i(1+i)^{44}3^{12}$	$-3^9 5^3 383^3/2^7$	I*30,II*
117*(1+i)3	o	o	o	-810i	972+972i	$-i(1+i)^{32}3^{12}$	$3^6 5^3/2$	I*18,II*
58*1+i	o	o	o	3030i	92212+92212i	$-i(1+i)^{72}3^4$	$-3^2 5^3 101^3/21$	I*58,II
56*i(1+i)	o	o	o	150i	-524-524i	$i(1+i)^{36}3^4$	$-3^2 5^6/2^3$	I*22,II

$$N=(1+i)^{12}(3)^4=(5184)$$

38*1	o	o	o	-15i	-16+16i	$-i(1+i)^{12}3^4$	$-2^6 3^2 5^3$	II ,II
38*i	o	o	o	15i	16+16i	$i(1+i)^{12}3^4$	$-2^6 3^2 5^3$	II ,II

38^*3	o	o	o	$-135i$	$-432+432i$	$-i(1+i)^{12}3^{10}$	$-2^63^25^3$	II ,IV*	•
38^*i3	o	o	o	$135i$	$432+432i$	$i(1+i)^{12}3^{10}$	$-2^63^25^3$	II ,IV*	•
38^*1+i	o	o	o	30	$-64i$	$-i(1+i)^{18}3^4$	$-2^63^25^3$	I*2 ,II	•
$38^*i(1+i)$	o	o	o	-30	-64	$i(1+i)^{18}3^4$	$-2^63^25^3$	I*2 ,II	•
$38^*(1+i)3$	o	o	o	270	$-1728i$	$-i(1+i)^{18}3^{10}$	$-2^63^25^3$	I*2 ,IV*	•
$38^*i(1+i)3$	o	o	o	-270	-1728	$i(1+i)^{18}3^{10}$	$-2^63^25^3$	I*2 ,IV*	•
37^*1	o	o	o	$3i$	$-2+2i$	$-i(1+i)^{12}3^4$	2^63^2	II ,II	•—• 13
39^*1	o	$3+3i$	o	$-12231i$	$-356187+356187i$	$-i(1+i)^{12}3^4$	$-2^63^24079^3$	II ,II	
37^*i	o	o	o	$-3i$	$2+2i$	$i(1+i)^{12}3^4$	2^63^2	II ,II	•—• 13
39^*i	o	$-3+3i$	o	$12231i$	$356187+356187i$	$i(1+i)^{12}3^4$	$-2^63^24079^3$	II ,II	
37^*3	o	o	o	$27i$	$-54+54i$	$-i(1+i)^{12}3^{10}$	2^63^2	II ,IV*	•—• 13
39^*3	o	$9+9i$	o	$-110079i$	$-9617049+9617049i$	$-i(1+i)^{12}3^{10}$	$-2^63^24079^3$	II ,IV*	
37^*i3	o	o	o	$-27i$	$54+54i$	$i(1+i)^{12}3^{10}$	2^63^2	II ,IV*	•—• 13
39^*i3	o	$-9+9i$	o	$110079i$	$9617049+9617049i$	$i(1+i)^{12}3^{10}$	$-2^63^24079^3$	II ,IV*	
37^*1+i	o	o	o	-6	$-8i$	$-i(1+i)^{18}3^4$	2^63^2	I*2 ,II	•—• 13
39^*1+i	o	$6i$	-6	24462	$-9-1424748i$	$-i(1+i)^{18}3^4$	$-2^63^24079^3$	I*2 ,II	

$37^{*}\,i(1+i)$	o	o	o	6	-8	$i(1+i)^{18}3^{4}$	$2^{6}3^{2}$	$I*2\ ,II$
$39^{*}\,i(1+i)$	o	-6	-6	-24462	-1424757	$i(1+i)^{18}3^{4}$	$-2^{6}3^{2}4079^{3}$	$I*2\ ,II$
$37^{*}\,(1+i)3$	o	o	o	-54	$-216i$	$-i(1+i)^{18}3^{10}$	$2^{6}3^{2}$	$I*2\ ,IV*$
$39^{*}\,(1+i)3$	o	$-9i$	-18	220239	$-81-40450347i$	$-i(1+i)^{18}3^{10}$	$-2^{6}3^{2}4079^{3}$	$I*2\ ,IV*$
$37^{*}\,i(1+i)3$	o	o	o	54	-216	$i(1+i)^{18}3^{10}$	$2^{6}3^{2}$	$I*2\ ,IV*$
$39^{*}\,i(1+i)3$	o	9	-18	-220239	-40450428	$i(1+i)^{18}3^{10}$	$-2^{6}3^{2}4079^{3}$	$I*2\ ,IV*$

$$N=(1+i)^{13}(3)^{4}=(5184+5184i)$$

$40^{*}1$	o	o	o	$-3i$	$-2+4i$	$-(1+i)^{13}3^{4}$	$i(1+i)^{11}3^{2}$	$II\ ,II$
$40^{*}i$	o	o	o	$3i$	$4+2i$	$(1+i)^{13}3^{4}$	$i(1+i)^{11}3^{2}$	$II\ ,II$
$40^{*}3$	o	o	o	$-27i$	$-54+108i$	$-(1+i)^{13}3^{10}$	$i(1+i)^{11}3^{2}$	$II\ ,IV*$
$40^{*}i3$	o	o	o	$27i$	$108+54i$	$(1+i)^{13}3^{10}$	$i(1+i)^{11}3^{2}$	$II\ ,IV*$
$40^{*}1+i$	o	o	o	6	$-4-12i$	$-(1+i)^{19}3^{4}$	$i(1+i)^{11}3^{2}$	$I*2\ ,II$
$40^{*}i(1+i)$	o	o	o	-6	$-12+4i$	$(1+i)^{19}3^{4}$	$i(1+i)^{11}3^{2}$	$I*2\ ,II$
$40^{*}(1+i)3$	o	o	o	54	$-108-324i$	$-(1+i)^{19}3^{10}$	$i(1+i)^{11}3^{2}$	$I*2\ ,IV*$

$40^*\,i(1+i)3$	o	o	o	-54	$-324+108i$	$(1+i)^{19}3^{10}$	$i(1+i)^{11}3^{2}$	I*2 ,IV*	•
41^*1	o	o	o	$-3i$	$-4+2i$	$-i(1+i)^{13}3^{4}$	$(1+i)^{11}3^{2}$	II ,II	•
41^*i	o	o	o	$3i$	$2+4i$	$i(1+i)^{13}3^{4}$	$(1+i)^{11}3^{2}$	II ,II	•
41^*3	o	o	o	$-27i$	$-108+54i$	$-i(1+i)^{13}3^{10}$	$(1+i)^{11}3^{2}$	II ,IV*	•
41^*i3	o	o	o	$27i$	$54+108i$	$i(1+i)^{13}3^{10}$	$(1+i)^{11}3^{2}$	II ,IV*	•
41^*1+i	o	o	o	6	$4-12i$	$-i(1+i)^{19}3^{4}$	$(1+i)^{11}3^{2}$	I*2 ,II	•
$41^*i(1+i)$	o	o	o	-6	$-12-4i$	$i(1+i)^{19}3^{4}$	$(1+i)^{11}3^{2}$	I*2 ,II	•
$41^*(1+i)3$	o	o	o	54	$108-324i$	$-i(1+i)^{19}3^{10}$	$(1+i)^{11}3^{2}$	I*2 ,IV*	•
$41^*i(1+i)3$	o	o	o	-54	$-324-108i$	$i(1+i)^{19}3^{10}$	$(1+i)^{11}3^{2}$	I*2 ,IV*	•
101^*1	o	o	o	$-9-27i$	$-18+54i$	$-i(1+i)^{14}3^{6}$	$-i2^{6}3^{3}(1+i)(2+i)^{3}$	III ,IV	•
101^*i	o	o	o	$9+27i$	$54+18i$	$i(1+i)^{14}3^{6}$	$-i2^{6}3^{3}(1+i)(2+i)^{3}$	III ,IV	•
101^*3	o	o	o	$-81-243i$	$-486+1458i$	$-i(1+i)^{14}3^{12}$	$-i2^{6}3^{3}(1+i)(2+i)^{3}$	III ,II*	•
101^*i3	o	o	o	$81+243i$	$1458+486i$	$i(1+i)^{14}3^{12}$	$-i2^{6}3^{3}(1+i)(2+i)^{3}$	III ,II*	•
101^*1+i	o	o	o	$54-18i$	$-72-144i$	$-i(1+i)^{20}3^{6}$	$-i2^{6}3^{3}(1+i)(2+i)^{3}$	III*,IV	•

101*i(1+i)	o	o	o	$-54+18i$	$-144+72i$	$i(1+i)^{20}3^{6}$	$-i2^{6}3^{3}(1+i)(2+i)^{3}$	III*,IV	●
101*(1+i)3	o	o	o	$486-162i$	$-1944-3888i$	$-i(1+i)^{20}3^{12}$	$-i2^{6}3^{3}(1+i)(2+i)^{3}$	III*,II*	●
101*i(1+i)3	o	o	o	$-486+162i$	$-3888+1944i$	$i(1+i)^{20}3^{12}$	$-i2^{6}3^{3}(1+i)(2+i)^{3}$	III*,II*	●
102*1	o	o	o	$-9+27i$	$-18-54i$	$-i(1+i)^{14}3^{6}$	$i2^{6}3^{3}(1+i)(1+2i)^{3}$	III ,IV	●
102*i	o	o	o	$9-27i$	$-54+18i$	$i(1+i)^{14}3^{6}$	$i2^{6}3^{3}(1+i)(1+2i)^{3}$	III ,IV	●
102*3	o	o	o	$-81+243i$	$-486-1458i$	$-i(1+i)^{14}3^{12}$	$i2^{6}3^{3}(1+i)(1+2i)^{3}$	III ,II*	●
102*i3	o	o	o	$81-243i$	$-1458+486i$	$i(1+i)^{14}3^{12}$	$i2^{6}3^{3}(1+i)(1+2i)^{3}$	III ,II*	●
102*1+i	o	o	o	$-54-18i$	$144+72i$	$-i(1+i)^{20}3^{6}$	$i2^{6}3^{3}(1+i)(1+2i)^{3}$	III*,IV	●
102*i(1+i)	o	o	o	$54+18i$	$72-144i$	$i(1+i)^{20}3^{6}$	$i2^{6}3^{3}(1+i)(1+2i)^{3}$	III*,IV	●
102*(1+i)3	o	o	o	$-486-162i$	$3888+1944i$	$-i(1+i)^{20}3^{12}$	$i2^{6}3^{3}(1+i)(1+2i)^{3}$	III*,II*	●
102*i(1+i)3	o	o	o	$486+162i$	$1944-3888i$	$i(1+i)^{20}3^{12}$	$i2^{6}3^{3}(1+i)(1+2i)^{3}$	III*,II*	●

$$N=(1+i)^{14}(3)^{4}=(10368)$$

$48^{*}1$	o	o	o	$3-9i$	$-12+4i$	$-(1+i)^{15}3^{4}$	$-i2^{6}3^{2}(2-i)^{3}$	III ,II	●
$48^{*}i$	o	o	o	$-3+9i$	$4+12i$	$(1+i)^{15}3^{4}$	$-i2^{6}3^{2}(2-i)^{3}$	III ,II	●
$48^{*}3$	o	o	o	$27-81i$	$-324+108i$	$-(1+i)^{15}3^{10}$	$-i2^{6}3^{2}(2-i)^{3}$	III ,IV*	●
$48^{*}i3$	o	o	o	$-27+81i$	$108+324i$	$(1+i)^{15}3^{10}$	$-i2^{6}3^{2}(2-i)^{3}$	III ,IV*	●
$48^{*}1+i$	o	o	o	$18+6i$	$16-32i$	$-(1+i)^{21}3^{4}$	$-i2^{6}3^{2}(2-i)^{3}$	III*,II	●
$48^{*}i(1+i)$	o	o	o	$-18-6i$	$-32-16i$	$(1+i)^{21}3^{4}$	$-i2^{6}3^{2}(2-i)^{3}$	III*,II	●
$48^{*}(1+i)3$	o	o	o	$162+54i$	$432-864i$	$-(1+i)^{21}3^{10}$	$-i2^{6}3^{2}(2-i)^{3}$	III*,IV*	●
$48^{*}i(1+i)3$	o	o	o	$-162-54i$	$-864-432i$	$(1+i)^{21}3^{10}$	$-i2^{6}3^{2}(2-i)^{3}$	III*,IV*	●
$49^{*}1$	o	o	o	$3+9i$	$-12-4i$	$-i(1+i)^{15}3^{4}$	$i2^{6}3^{2}(2+i)^{3}$	III ,II	●
$49^{*}i$	o	o	o	$-3-9i$	$-4+12i$	$i(1+i)^{15}3^{4}$	$i2^{6}3^{2}(2+i)^{3}$	III ,II	●
$49^{*}3$	o	o	o	$27+81i$	$-324-108i$	$-i(1+i)^{15}3^{10}$	$i2^{6}3^{2}(2+i)^{3}$	III ,IV*	●
$49^{*}i3$	o	o	o	$-27-81i$	$-108+324i$	$i(1+i)^{15}3^{10}$	$i2^{6}3^{2}(2+i)^{3}$	III ,IV*	●
$49^{*}1+i$	o	o	o	$-18+6i$	$32-16i$	$-i(1+i)^{21}3^{4}$	$i2^{6}3^{2}(2+i)^{3}$	III*,II	●

$49^*i(1+i)$	O	O	O	$18-6i$	$-16-32i$	$i(1+i)^{21}3^4$	$i2^63^2(2+i)^3$	III*,II	●
$49^*(1+i)3$	O	O	O	$-162+54i$	$864-432i$	$-i(1+i)^{21}3^{10}$	$i2^63^2(2+i)^3$	III*,IV*	●
$49^*i(1+i)3$	O	O	O	$162-54i$	$-432-864i$	$i(1+i)^{21}3^{10}$	$i2^63^2(2+i)^3$	III*,IV*	●

$$N=(3)^5=(243)$$

60^*1	O	O	$-i$	O	1	-3^5	O	IO ,II	
60^*3	O	O	-1	O	20	-3^{11}	O	IO ,IV*	
121^*1	O	O	-1	O	2	-3^7	O	IO ,IV	
121^*3	O	O	$-i$	O	61	-3^{13}	O	IO ,II*	

$$N=(1+i)(3)^5=(243+243i)$$

87^*1	$-i$	1	$-i$	-10	11	$-i(1+i)^23^5$	$3^419^3/2$	I2 ,II	
89^*i3	-1	-1	-1	-56	163	$i(1+i)^63^{11}$	$3^411^3/2^3$	I6 ,IV*	
87^*3	-1	-1	O	-96	386	$-i(1+i)^23^{11}$	$3^419^3/2$	I2 ,IV*	
89^*i	$-i$	1	O	-6	4	$i(1+i)^63^5$	$3^411^3/2^3$	I6 ,II	

88^*i	$-i$	1	$-i$	-19	29	$(1+i)^4 3^5$	$-3^4 5^3 7^3/2^3$	I4 ,II
92^*i3	$-i$	1	$-i$	26	161	$(1+i)^{12}3^{11}$	$3^4 5^3/2^6$	I12 ,IV*
88^*i3	-1	-1	O	-177	953	$(1+i)^4 3^{11}$	$-3^4 5^3 7^3/2^2$	I4 ,IV*
92^*i	-1	-1	O	3	5	$(1+i)^{12}3^5$	$3^4 5^3/2^6$	I12 ,II
141^*i	-1	-1	O	-123	557	$i(1+i)^6 3^7$	$3^5 73^3/2^3$	I6 ,IV
142^*3	$-i$	1	O	-258	748	$-i(1+i)^{18}3^{13}$	$3^5 17^3/2^9$	I18 ,II*
141^*i3	$-i$	1	$-i$	-1108	13931	$i(1+i)^6 3^{13}$	$3^5 73^3/2^3$	I6 ,II*
142^*1	-1	-1	-1	-29	37	$-i(1+i)^{18}3^7$	$3^5 17^3/2^9$	I18 ,IV

$$N=(1+i)^2(3)^5=(486)$$

66^*1+i	O	O	O	O	3	$-(1+i)^8 3^5$	O	IV* ,II
$66^*(1+i)3$	O	O	O	O	81	$-(1+i)^8 3^{11}$	O	IV* ,IV*
76^*1	O	O	$-1-i$	O	$-2i$	$-(1+i)^4 3^5$	O	IV ,II
76^*3	O	O	$-1-i$	O	$-41i$	$-(1+i)^4 3^{11}$	O	IV ,IV*
$130^*(1+i)$	O	O	O	O	9	$-(1+i)^8 3^7$	O	IV* ,IV
$130^*(1+i)3$	O	O	O	O	243	$-(1+i)^8 3^{13}$	O	IV* ,II*
135^*1	O	O	$-1-i$	O	$-5i$	$-(1+i)^4 3^7$	O	IV ,IV
135^*3	O	O	$-1-i$	O	$-122i$	$-(1+i)^4 3^{13}$	O	IV ,II*

$$N=(1+i)^3 (3)^5 =(486+486i)$$

75^*1	$-1-i$	i	0	-3	$-2i$	$-(1+i)^4 3^5$	$2^4 3^4$	III ,II	●
75^*3	$-1-i$	i	$-1-i$	$-21-i$	$-31i$	$-(1+i)^4 3^{11}$	$2^4 3^4$	III ,IV*	●
78^*1	0	0	$-1-i$	9	$-11i$	$(1+i)^4 3^5$	$-2^{10} 3^4$	III ,II	●
78^*3	0	0	$-1-i$	81	$-284i$	$-(1+i)^4 3^{11}$	$-2^{10} 3^4$	III ,IV*	●
86^*i	$-1-i$	i	0	6	$-2i$	$i(1+i)^{10} 3^5$	$2 \cdot 3^7$	III*,II	●
86^*i3	$-1-i$	i	$-1-i$	$60-i$	$-112i$	$i(1+i)^{10} 3^{11}$	$2 \cdot 3^7$	III*,IV*	●
131^*1+i	0	0	0	-54	153	$-(1+i)^8 3^7$	$-2^{11} 3^5$	I*1 ,IV	●
$131^*(1+i)3$	0	0	0	-486	4131	$-(1+i)^8 3^{13}$	$-2^{11} 3^5$	I*1 ,II*	●
139^*i	$-1-i$	i	0	6	$-8i$	$(1+i)^8 3^7$	$-2^2 3^5$	I*1 ,IV	●
139^*i3	$-1-i$	i	$-1-i$	$60-i$	$-274i$	$(1+i)^8 3^{13}$	$-2^2 3^5$	I*1 ,II*	●

$$N=(1+i)^6(3)^5=(1944)$$

$61^{*}i$	o	o	o	-9	-12	$(1+i)^{12}3^5$	$-2^6 3^4$	I*2 ,II	●
$61^{*}i3$	o	o	o	-81	-324	$(1+i)^{12}3^{11}$	$-2^6 3^4$	I*2 ,IV*	●
$62^{*}1$	o	o	o	18	$-30i$	$-(1+i)^{12}3^5$	$-2^9 3^4$	I*2 ,II	●
$62^{*}3$	o	o	o	162	$-810i$	$-(1+i)^{12}3^{11}$	$-2^9 3^4$	I*2 ,IV*	●
$81^{*}1$	$-1-i$	i	o	$-3-9i$	$9-11i$	$-i(1+i)^6 3^5$	$i2^3 3^4(4-i)^3$	II ,II	●
$81^{*}3$	$-1-i$	i	$-1-i$	$-21-82i$	$162-274i$	$-i(1+i)^6 3^{11}$	$i2^3 3^4(4-i)^3$	II ,IV*	●
$82^{*}1$	$-1-i$	i	o	$-3+9i$	$-9-11i$	$-i(1+i)^6 3^5$	$-i2^3 3^4(4+i)^3$	II ,II	●
$82^{*}3$	$-1-i$	i	$-1-i$	$-21+80i$	$-162-274i$	$-i(1+i)^6 3^{11}$	$-i2^3 3^4(4+i)^3$	II ,IV*	●
$136^{*}i$	$-1-i$	i	o	6	i	$-i(1+i)^6 3^7$	$2^3 3^5$	II ,IV	●
$136^{*}i3$	$-1-i$	i	$-1-i$	$60-i$	$-31i$	$-i(1+i)^6 3^{13}$	$2^3 3^5$	II ,II*	●
$137^{*}1$	$-1-i$	i	o	1356	$-18557i$	$-i(1+i)^6 3^7$	$2^3 3^8 67^3$	II ,IV	●
$137^{*}3$	$-1-i$	i	$-1-i$	$12210-i$	$-513247i$	$-i(1+i)^6 3^{13}$	$2^3 3^8 67^3$	II ,II*	●

$$N=(1+i)^7(3)^5=(1944+1944i)$$

79^*i	o	o	o	-9	$-12+6i$	$i(1+i)^{16}3^5$	$-i2^43^4$	I*5 ,II	●
79^*i3	o	o	o	-81	$-324+162i$	$i(1+i)^{16}3^{11}$	$-i2^43^4$	I*5 ,IV*	●
80^*1	o	o	o	-9	$-12-6i$	$-i(1+i)^{16}3^5$	$i2^43^4$	I*5 ,II	●
80^*3	o	o	o	-81	$-324-162i$	$-i(1+i)^{16}3^{11}$	$i2^43^4$	I*5 ,IV*	●
132^*1	o	o	o	$27-54i$	$-180+18i$	$-(1+i)^{14}3^7$	$-2^53^5(2+i)^3$	III*,IV	●
132^*3	o	o	o	$243-486i$	$-4860+486i$	$-(1+i)^{14}3^{13}$	$-2^53^5(2+i)^3$	III*,II*	●
133^*i	o	o	o	$27+54i$	$180+18i$	$(1+i)^{14}3^7$	$-2^53^5(2-i)^3$	III*,IV	●
133^*i3	o	o	o	$243+486i$	$4860+486i$	$(1+i)^{14}3^{13}$	$-2^53^5(2-i)^3$	III*,II*	●

$$N=(1+i)^8(3)^5=(3888)$$

61^*1	o	o	o	9	$-12i$	$-(1+i)^{12}3^5$	-2^63^4	I*O ,II	●
61^*3	o	o	o	81	$-324i$	$-(1+i)^{12}3^{11}$	-2^63^4	I*O ,IV*	●
62^*i	o	o	o	-18	-30	$(1+i)^{12}3^5$	-2^93^4	I*O ,II	●

$62^{*}i3$	o	o	o	-162	-810	$(1+i)^{12}3^{11}$	$-2^{9}5^{4}$	I*O ,IV*	●
$75^{*}i$	o	o	o	-9	6i	$(1+i)^{16}3^{5}$	$2^{4}3^{4}$	I*4 ,II	●
$75^{*}i3$	o	o	o	-81	162i	$(1+i)^{16}3^{11}$	$2^{4}3^{4}$	I*4 ,IV*	●
$78^{*}i$	o	o	o	36	84i	$(1+i)^{16}3^{5}$	$-2^{10}3^{4}$	II* ,II	●
$78^{*}i3$	o	o	o	324	2268i	$(1+i)^{16}3^{11}$	$-2^{10}3^{4}$	II* ,IV*	●
$79^{*}1$	o	o	o	9	-6-12i	$-i(1+i)^{16}3^{5}$	$-i2^{4}3^{4}$	II* ,II	●
$79^{*}3$	o	o	o	81	-162-324i	$-i(1+i)^{16}3^{11}$	$-i2^{4}3^{4}$	II* ,IV*	●
$80^{*}i$	o	o	o	9	-6+12i	$i(1+i)^{16}3^{5}$	$i2^{4}3^{4}$	II* ,II	●
$80^{*}i3$	o	o	o	81	-162+324i	$i(1+i)^{16}3^{11}$	$i2^{4}3^{4}$	II* ,IV*	●
$81^{*}i$	o	o	o	-9-36i	-36+78i	$i(1+i)^{18}3^{5}$	$i2^{3}3^{4}(4-i)^{3}$	I*6 ,II	●
$81^{*}i3$	o	o	o	-81-324i	-972+2106i	$i(1+i)^{18}3^{11}$	$i2^{3}3^{4}(4-i)^{3}$	I*6 ,IV*	●
$82^{*}i$	o	o	o	-9+36i	36+78i	$i(1+i)^{18}3^{5}$	$-i2^{3}3^{4}(4+i)^{3}$	I*6 ,II	●
$82^{*}i3$	o	o	o	-81+324i	972+2106i	$i(1+i)^{18}3^{11}$	$-i2^{3}3^{4}(4+i)^{3}$	I*6 ,IV*	●
$86^{*}1$	o	o	o	27	-42i	$-i(1+i)^{22}3^{5}$	$2\cdot3^{7}$	I*10,II	●

86*3	o	o	o	243	-1134i	$-i(1+i)^{22}3^{11}$	$2\cdot3^7$	I*10,IV* •
131*i(1+i)	o	o	o	54	-153i	$(1+i)^8 3^7$	$-2^{11}3^5$	II ,IV •
131*i(1+i)3	o	o	o	486	-4131i	$(1+i)^8 3^{13}$	$-2^{11}3^5$	II ,II* •
132*i	o	o	o	-27+54i	18+180i	$(1+i)^{14}3^7$	$-2^5 3^5(2+i)^3$	I*2 ,IV •
132*i3	o	o	o	-243+486i	486+4860i	$(1+i)^{14}3^{13}$	$-2^5 3^5(2+i)^3$	I*2 ,II* •
133*1	o	o	o	-27-54i	-18+180i	$-(1+i)^{14}3^7$	$-2^5 3^5(2-i)^3$	I*2 ,IV •
133*3	o	o	o	-243-486i	-486+4860i	$-(1+i)^{14}3^{13}$	$-2^5 3^5(2-i)^3$	I*2 ,II* •
136*1	o	o	o	27	-18i	$i(1+i)^{18}3^7$	$2^3 3^5$	I*6 ,IV •
136*3	o	o	o	243	-486i	$i(1+i)^{18}3^{13}$	$2^3 3^5$	I*6 ,II* •
137*i	2i	-8	-6-12i	5442+6i	-16281+153846i	$i(1+i)^{18}3^7$	$2^3 3^8 67^3$	I*6 ,IV •
137*i3	6i	9	o	48843	4154814i	$i(1+i)^{18}3^{13}$	$2^3 3^8 67^3$	I*6 ,II* •
139*1	o	o	o	27	-90i	$-(1+i)^{20}3^7$	$-2^2 3^5$	I*8 ,IV •
139*3	o	o	o	243	-2430i	$-(1+i)^{20}3^{13}$	$-2^2 3^5$	I*8 ,II* •

$60^{*}i$	o	o	o	o	-6	$(1+i){}^{12}_{3}{}^{5}$	o	I*O ,II	•—• 3
$60^{*}i3$	o	o	o	o	-162	$(1+i){}^{12}_{3}{}^{11}$	o	I*O ,IV*	
$66^{*}i(1+i)$	o	o	o	o	-3i	$(1+i){}^{8}_{3}{}^{5}$	o	II ,II	•—• 3
$66^{*}i(1+i)3$	o	o	o	o	-81i	$(1+i){}^{8}_{3}{}^{11}$	o	II ,IV*	
$76^{*}i$	o	o	o	o	12i	$(1+i){}^{16}_{3}{}^{5}$	o	II* ,II	•—• 3
$76^{*}i3$	o	o	o	o	324i	$(1+i){}^{16}_{3}{}^{11}$	o	II* ,IV*	
$87^{*}i3$	o	o	o	1539	23166i	$i(1+i){}^{26}_{3}{}^{11}$	$3^4 19^3/2$	I*14,IV*	•—• 3
$89^{*}1$	o	o	o	99	-354i	$-i(1+i){}^{30}_{3}{}^{5}$	$3^4 11^3/2^3$	I*18,II	
$87^{*}i$	o	o	o	171	858i	$i(1+i){}^{26}_{3}{}^{5}$	$3^4 19^3/2$	I*14,II	•—• 3
$89^{*}3$	o	o	o	891	-9558i	$-i(1+i){}^{30}_{3}{}^{11}$	$3^4 11^3/2^3$	I*18,IV*	
$88^{*}1$	2i	1-9i	66i	354	1089-3072i	$-(1+i){}^{28}_{3}{}^{5}$	$-3^4 5^3 7^3/2^2$	I*16,II	•—• 3
$92^{*}3$	o	o	o	-405	-9882i	$-(1+i){}^{36}_{3}{}^{11}$	$3^4 5^3/2^6$	I*24,IV*	
$88^{*}3$	6i	9-27i	198i	3186	9801-82944i	$-(1+i){}^{28}_{3}{}^{11}$	$-3^4 5^3 7^3/2^2$	I*16,IV*	•—• 3
$92^{*}1$	o	o	o	-45	-366i	$-(1+i){}^{36}_{3}{}^{5}$	$3^4 5^3/2^6$	I*24,II	
$121^{*}i$	o	o	o	o	-18	$(1+i){}^{12}_{3}{}^{7}$	o	I*O ,IV	•—• 3
$121^{*}i3$	o	o	o	o	-486	$(1+i){}^{12}_{3}{}^{13}$	o	I*O ,II*	
$130^{*}i(1+i)$	o	o	o	o	-9i	$(1+i){}^{8}_{3}{}^{7}$	o	II ,IV	•—• 3
$130^{*}i(1+i)3$	o	o	o	o	-243i	$(1+i){}^{8}_{3}{}^{13}$	o	II ,II*	

- 186 -

$135^{*}i$	0	0	0	0	36i	$(1+i)^{16}3^{7}$	0	II* ,IV	●—● 3
$135^{*}i3$	0	0	0	0	972i	$(1+i)^{16}3^{13}$	0	II* ,II*	
$141^{*}1$	2i	1−27i	30+30i	1758−30i	−51138i	$-i(1+i)^{30}3^{7}$	$3^{5}73^{3}/2^{3}$	I*18,IV	●—● 3
$142^{*}i3$	6i	9−135i	756+2322i	5022−2268i	1205037−920484i	$i(1+i)^{42}3^{13}$	$3^{5}17^{3}/2^{9}$	I*30,II*	
$141^{*}3$	6i	9+108i	−108+108i	14175+324i	−311526i	$-i(1+i)^{30}3^{13}$	$3^{5}73^{3}/2^{3}$	I*18,II*	●—● 3
$142^{*}i$	2i	766+207i	258+1278i	182529+105312i	13447935+13000401i	$i(1+i)^{42}3^{7}$	$3^{5}17^{3}/2^{9}$	I*30,IV	

$$N=(1+i)^{9}(3)^{5}=(3888+3888i)$$

$72^{*}1+i$	0	0	0	18i	12+12i	$-i(1+i)^{20}3^{5}$	$2^{5}3^{4}$	I*7 ,II	●
$72^{*}i(1+i)$	0	0	0	−18i	12−12i	$i(1+i)^{20}3^{5}$	$2^{5}3^{4}$	I*7 ,II	●
$72^{*}(1+i)3$	0	0	0	162i	324+324i	$-i(1+i)^{20}3^{11}$	$2^{5}3^{4}$	I*7 ,IV*	●
$72^{*}i(1+i)3$	0	0	0	−162i	324−324i	$i(1+i)^{20}3^{11}$	$2^{5}3^{4}$	I*7 ,IV*	●

$$N=(1+i)^{10}(3)^{5}=(7776)$$

$61^{*}1+i$	0	0	0	18i	24+24i	$-(1+i)^{18}3^{5}$	$-2^{6}3^{4}$	I*4 ,II	●

$61^*i(1+i)$	o	o	o	$-18i$	$24-24i$	$(1+i)^{18}3^5$	$-2^6 3^4$	I*4 ,II	●
$61^*(1+i)3$	o	o	o	$162i$	$648+648i$	$-(1+i)^{18}3^{11}$	$-2^6 3^4$	I*4 ,IV*	●
$61^*i(1+i)3$	o	o	o	$-162i$	$648-648i$	$(1+i)^{18}3^{11}$	$-2^6 3^4$	I*4 ,IV*	●
62^*1+i	o	o	o	$36i$	$60+60i$	$-(1+i)^{18}3^5$	$-2^9 3^4$	II* ,II	●
$62^*i(1+i)$	o	o	o	$-36i$	$60-60i$	$(1+i)^{18}3^5$	$-2^9 3^4$	II* ,II	●
$62^*(1+i)3$	o	o	o	$324i$	$1620+1620i$	$-(1+i)^{18}3^{11}$	$-2^9 3^4$	II* ,IV*	●
$62^*i(1+i)3$	o	o	o	$-324i$	$1620-1620i$	$(1+i)^{18}3^{11}$	$-2^9 3^4$	II* ,IV*	●
72^*1	o	o	o	9	$-6i$	$-i(1+i)^{14}3^5$	$2^5 3^4$	I*O ,II	●
72^*i	o	o	o	-9	-6	$i(1+i)^{14}3^5$	$2^5 3^4$	I*O ,II	●
72^*3	o	o	o	81	$-162i$	$-i(1+i)^{14}3^{11}$	$2^5 3^4$	I*O ,IV*	●
72^*i3	o	o	o	-81	-162	$i(1+i)^{14}3^{11}$	$2^5 3^4$	I*O ,IV*	●
75^*1+i	o	o	o	$18i$	$12-12i$	$-(1+i)^{22}3^5$	$2^4 3^5$	I*8 ,II	●
$75^*i(1+i)$	o	o	o	$-18i$	$-12-12i$	$(1+i)^{22}3^5$	$2^4 3^4$	I*8 ,II	●
$75^*(1+i)3$	o	o	o	$162i$	$324-324i$	$-(1+i)^{22}3^{11}$	$2^4 3^4$	I*8 ,IV*	●

75 *i(1+i)3	O	O	O	$-162i$	$-324-324i$	$(1+i)^{22}3^{11}$	$2^4 3^4$	I*8 ,IV*	•
78 *1+i	O	O	O	$18i$	$21+21i$	$-(1+i)^{10}3^5$	$-2^{10}3^4$	II ,II	•
78 *i(1+i)	O	O	O	$-18i$	$21-21i$	$(1+i)^{10}3^5$	$-2^{10}3^4$	II ,II	•
78 *(1+i)3	O	O	O	$162i$	$567+567i$	$-(1+i)^{10}3^{11}$	$-2^{10}3^4$	II ,IV*	•
78 *i(1+i)3	O	O	O	$-162i$	$567-567i$	$(1+i)^{10}3^{11}$	$-2^{10}3^4$	II ,IV*	•
79 *1+i	O	O	O	$18i$	$36+12i$	$-i(1+i)^{22}3^5$	$-i2^4 3^4$	I*8 ,II	•
79 *i(1+i)	O	O	O	$-18i$	$12-36i$	$i(1+i)^{22}3^5$	$-i2^4 3^4$	I*8 ,II	•
79 *(1+i)3	O	O	O	$162i$	$972+324i$	$-i(1+i)^{22}3^{11}$	$-i2^4 3^4$	I*8 ,IV*	•
79 *i(1+i)3	O	O	O	$-162i$	$324-972i$	$i(1+i)^{22}3^{11}$	$-i2^4 3^4$	I*8 ,IV*	•
80 *1+i	O	O	O	$-18i$	$36-12i$	$-i(1+i)^{22}3^5$	$i2^4 3^4$	I*8 ,II	•
80 *i(1+i)	O	O	O	$18i$	$-12-36i$	$i(1+i)^{22}3^5$	$i2^4 3^4$	I*8 ,II	•
80 *(1+i)3	O	O	O	$-162i$	$972-324i$	$-i(1+i)^{22}3^{11}$	$i2^4 3^4$	I*8 ,IV*	•
80 *i(1+i)3	O	O	O	$162i$	$-324-972i$	$i(1+i)^{22}3^{11}$	$i2^4 3^4$	I*8 ,IV*	•
81 *1+i	O	O	O	$-72+18i$	$228-84i$	$-i(1+i)^{24}3^5$	$i2^3 3^4 (4-i)^3$	I*10,II	•

$81^* i(1+i)$	o	o	o	$72-18i$	$-84-228i$	$i(1+i)^{24}3^5$	$i2^3 3^4(4-i)^3$	I*10,II	●
$81^* (1+i)3$	o	o	o	$-648+162i$	$6156-2268i$	$-i(1+i)^{24}3^{11}$	$i2^3 3^4(4-i)^3$	I*10,IV*	●
$81^* i(1+i)3$	o	o	o	$648-162i$	$-2268-6156i$	$i(1+i)^{24}3^{11}$	$i2^3 3^4(4-i)^3$	I*10,IV*	●
$82^* 1+i$	o	o	o	$72+18i$	$84-228i$	$-i(1+i)^{24}3^5$	$-i2^3 3^4(4+i)^3$	I*10,II	●
$82^* i(1+i)$	o	o	o	$-72-18i$	$-228-84i$	$i(1+i)^{24}3^5$	$-i2^3 3^4(4+i)^3$	I*10,II	●
$82^* (1+i)3$	o	o	o	$648+162i$	$2268-6156i$	$-i(1+i)^{24}3^{11}$	$-i2^3 3^4(4+i)^3$	I*10,IV*	●
$82^* i(1+i)3$	o	o	o	$-648-162i$	$-6156-2268i$	$i(1+i)^{24}3^{11}$	$-i2^3 3^4(4+i)^3$	I*10,IV*	●
$86^* 1+i$	o	o	o	$54i$	$84+84i$	$-i(1+i)^{28}3^5$	$2\cdot3^7$	I*14,II	●
$86^* i(1+i)$	o	o	o	$-54i$	$84-84i$	$i(1+i)^{28}3^5$	$2\cdot3^7$	I*14,II	●
$86^* (1+i)3$	o	o	o	$486i$	$2268+2268i$	$-i(1+i)^{28}3^{11}$	$2\cdot3^7$	I*14,IV*	●
$86^* i(1+i)3$	o	o	o	$-486i$	$2268-2268i$	$i(1+i)^{28}3^{11}$	$2\cdot3^7$	I*14,IV*	●
$131^* 1$	o	o	o	$-108i$	$-306+306i$	$-(1+i)^{14}3^7$	$-2^{11}3^5$	I*0 ,IV	●
$131^* i$	o	o	o	$108i$	$306+306i$	$(1+i)^{14}3^7$	$-2^{11}3^5$	I*0 ,IV	●
$131^* 3$	o	o	o	$-972i$	$-8262+8262i$	$-(1+i)^{14}3^{13}$	$-2^{11}3^5$	I*0 ,II*	●

$131^*\,i3$	O	O	O	$972i$	$8262+8262i$	$(1+i)^{14}3^{13}$	$-2^{11}3^5$	I*O ,II*	●
$132^*\,1+i$	O	O	O	$108+54i$	$324-396i$	$-(1+i)^{20}3^7$	$-2^5 3^5(2+i)^3$	I*6 ,IV	●
$132^*\,i(1+i)$	O	O	O	$-108-54i$	$-396-324i$	$(1+i)^{20}3^7$	$-2^5 3^5(2+i)^3$	I*6 ,IV	●
$132^*\,(1+i)3$	O	O	O	$972+486i$	$8748-10692i$	$-(1+i)^{20}3^{13}$	$-2^5 3^5(2+i)^3$	I*6 ,II*	●
$132^*\,i(1+i)3$	O	O	O	$-972-486i$	$-10692-8748i$	$(1+i)^{20}3^{13}$	$-2^5 3^5(2+i)^3$	I*6 ,II*	●
$133^*\,1+i$	O	O	O	$108-54i$	$-324-396i$	$-(1+i)^{20}3^7$	$-2^5 3^5(2-i)^3$	I*6 ,IV	●
$133^*\,i(1+i)$	O	O	O	$-108+54i$	$-396+324i$	$(1+i)^{20}3^7$	$-2^5 3^5(2-i)^3$	I*6 ,IV	●
$133^*\,(1+i)3$	O	O	O	$972-486i$	$-8748-10692i$	$-(1+i)^{20}3^{13}$	$-2^5 3^5(2-i)^3$	I*6 ,II*	●
$133^*\,i(1+i)3$	O	O	O	$-972+486i$	$-10692+8748i$	$(1+i)^{20}3^{13}$	$-2^5 3^5(2-i)^3$	I*6 ,II*	●
$136^*\,1+i$	O	O	O	$54i$	$36+36i$	$i(1+i)^{24}3^7$	$2^3 3^5$	I*10,IV	●
$136^*\,i(1+i)$	O	O	O	$-54i$	$36-36i$	$-i(1+i)^{24}3^7$	$2^3 3^5$	I*10,IV	●
$136^*\,(1+i)3$	O	O	O	$486i$	$972+972i$	$i(1+i)^{24}3^{13}$	$2^3 3^5$	I*10,II*	●
$136^*\,i(1+i)3$	O	O	O	$-486i$	$972-972i$	$-i(1+i)^{24}3^{13}$	$2^3 3^5$	I*10,II*	●
$137^*\,1+i$	O	$-18i$	$24-6i$	$-108-10854i$	$242505-307476i$	$-i(1+i)^{24}3^7$	$2^3 3^8 67^3$	I*10,IV	●

$137^{*}i(1+i)$	0	$-18i$	$24-6i$	$-108+10854i$	$-242775-307476i$	$i(1+i)^{24}3^{7}$	$2^{3}3^{8}67^{3}$	I*10,IV	•
$137^{*}(1+i)3$	0	0	0	$-97686i$	$8309628-8309628i$	$-i(1+i)^{24}3^{13}$	$2^{3}3^{8}67^{3}$	I*10,II*	•
$137^{*}i(1+i)3$	0	$-27i$	0	$-243+97686i$	$-7430454-8308899i$	$i(1+i)^{24}3^{13}$	$2^{3}3^{8}67^{3}$	I*10,II*	•
$139^{*}1+i$	0	0	0	$54i$	$180+180i$	$-(1+i)^{26}3^{7}$	$-2^{2}3^{5}$	I*12,IV	•
$139^{*}i(1+i)$	0	0	0	$-54i$	$180-180i$	$(1+i)^{26}3^{7}$	$-2^{2}3^{5}$	I*12,IV	•
$139^{*}(1+i)3$	0	0	0	$486i$	$4860+4860i$	$-(1+i)^{26}3^{13}$	$-2^{2}3^{5}$	I*12,II*	•
$139^{*}i(1+i)3$	0	0	0	$-486i$	$4860-4860i$	$(1+i)^{26}3^{13}$	$-2^{2}3^{5}$	I*12,II*	•
$60^{*}1+i$	0	0	0	0	$12+12i$	$-(1+i)^{18}3^{5}$	0	II* ,II	•—• 3
$60^{*}(1+i)3$	0	0	0	0	$324+324i$	$-(1+i)^{18}3^{11}$	0	II* ,IV*	
$60^{*}i(1+i)$	0	0	0	0	$12-12i$	$(1+i)^{18}3^{5}$	0	II* ,II	•—• 3
$60^{*}i(1+i)3$	0	0	0	0	$324-324i$	$(1+i)^{18}3^{11}$	0	II* ,IV*	
$66^{*}1$	0	0	0	0	$-6+6i$	$-(1+i)^{14}3^{5}$	0	I*O ,II	•—• 3
$66^{*}3$	0	0	0	0	$-162+162i$	$-(1+i)^{14}3^{11}$	0	I*O ,IV*	
$66^{*}i$	0	0	0	0	$6+6i$	$(1+i)^{14}3^{5}$	0	I*O ,II	•—• 3
$66^{*}i3$	0	0	0	0	$162+162i$	$(1+i)^{14}3^{11}$	0	I*O ,IV*	

76*1+i	O	O	O	O	3+3i	$-(1+i)^{10}3^5$	O	II ,II	●—●³
76*(1+i)3	O	O	O	O	81+81i	$-(1+i)^{10}3^{11}$	O	II ,IV*	
76*i(1+i)	O	O	O	O	3-3i	$(1+i)^{10}3^5$	O	II ,II	●—●³
76*i(1+i)3	O	O	O	O	81-81i	$(1+i)^{10}3^{11}$	O	II ,IV*	
87*1+i	O	O	O	-342i	1716-1716i	$-i(1+i)^{32}3^5$	$3^4 19^3/2$	I*18,II	●—●³
89*i(1+i)3	O	O	O	-1782i	19116-19116i	$i(1+i)^{36}3^{11}$	$3^4 11^3/2^3$	I*22,IV*	
87*i(1+i)	O	O	O	342i	-1716-1716i	$i(1+i)^{32}3^5$	$3^4 19^3/2$	I*18,II	●—●³
89*(1+i)3	O	O	O	1782i	19116+19116i	$-i(1+i)^{36}3^{11}$	$3^4 11^3/2^3$	I*22,IV*	
87*(1+i)3	O	O	O	-3078i	46332-46332i	$-i(1+i)^{32}3^{11}$	$3^4 19^3/2$	I*18,IV*	●—●³
89*i(1+i)	O	O	O	-198i	708-708i	$i(1+i)^{36}3^5$	$3^4 11^3/2^3$	I*22,II	
87*i(1+i)3	O	O	O	3078i	-46332-46332i	$i(1+i)^{32}3^{11}$	$3^4 19^3/2$	I*18,IV*	●—●³
89*1+i	O	O	O	198i	708+708i	$-i(1+i)^{36}3^5$	$3^4 11^3/2^3$	I*22,II	
88*1+i	O	9-9i	O	576i	6144+6144i	$-(1+i)^{34}3^5$	$-3^4 5^3 7^3/2^2$	I*20,II	●—●³
92*(1+i)3	O	O	O	-810i	19764+19764i	$-(1+i)^{42}3^{11}$	$3^4 5^3/2^6$	I*28,IV*	
88*i(1+i)	O	9+9i	O	-576i	6144-6144i	$-(1+i)^{34}3^5$	$-3^4 5^3 7^3/2^2$	I*20,II	●—●³
92*i(1+i)3	O	O	O	810i	19764-19764i	$-(1+i)^{42}3^{11}$	$3^4 5^3/2^6$	I*28,IV*	
88*(1+i)3	O	27-27i	O	5184i	165888+165888i	$-(1+i)^{34}3^{11}$	$-3^4 5^3 7^3/2^2$	I*20,IV*	●—●³
92*1+i	O	O	O	-90i	732+732i	$-(1+i)^{42}3^5$	$3^4 5^3/2^6$	I*28,II	

88*i(1+i)3	0	$27+27i$	0	$-5184i$	$165888-165888i$	$(1+i)^{34}3^{11}$	$-3^4 5^3 7^3/2^2$	I*20,IV*	$\bullet\!-\!\bullet\,{}^{3}$
92*i(1+i)	0	0	0	$90i$	$732-732i$	$(1+i)^{42}3^{5}$	$3^4 5^3/2^6$	I*28,II	
121*1+i	0	0	0	0	$36+36i$	$-(1+i)^{18}3^{7}$	0	II* ,IV	$\bullet\!-\!\bullet\,{}^{3}$
121*(1+i)3	0	0	0	0	$972+972i$	$-(1+i)^{18}3^{13}$	0	II* ,II*	
121*i(1+i)	0	0	0	0	$36-36i$	$(1+i)^{18}3^{7}$	0	II* ,IV	$\bullet\!-\!\bullet\,{}^{3}$
121*i(1+i)3	0	0	0	0	$972-972i$	$(1+i)^{18}3^{13}$	0	II* ,II*	
130*1	0	0	0	0	$-18+18i$	$-(1+i)^{14}3^{7}$	0	I*0 ,IV	$\bullet\!-\!\bullet\,{}^{3}$
130*3	0	0	0	0	$-486+486i$	$-(1+i)^{14}3^{13}$	0	I*0 ,II*	
130*i	0	0	0	0	$18+18i$	$(1+i)^{14}3^{7}$	0	I*0 ,IV	$\bullet\!-\!\bullet\,{}^{3}$
130*i3	0	0	0	0	$486+486i$	$(1+i)^{14}3^{13}$	0	I*0 ,II*	
135*1+i	0	0	0	0	$9+9i$	$-(1+i)^{10}3^{7}$	0	II ,IV	$\bullet\!-\!\bullet\,{}^{3}$
135*(1+i)3	0	0	0	0	$243+243i$	$-(1+i)^{10}3^{13}$	0	II ,II*	
135*i(1+i)	0	0	0	0	$9-9i$	$(1+i)^{10}3^{7}$	0	II ,IV	$\bullet\!-\!\bullet\,{}^{3}$
135*i(1+i)3	0	0	0	0	$243-243i$	$(1+i)^{10}3^{13}$	0	II ,II*	
141*1+i	0	$-162-126i$	$-66-66i$	$3456+17550i$	$361224-441018i$	$-i(1+i)^{36}3^{7}$	$3^5 73^3/2^3$	I*22,IV	$\bullet\!-\!\bullet\,{}^{3}$
142*i(1+i)3	0	$-621+1389i$	$-1536+512i$	$-514560-566784i$	$119799808-41156608i$	$i(1+i)^{48}3^{13}$	$3^5 17^3/2^9$	I*34,II*	
141*i(1+i)	0	$-162-72i$	$-66-192i$	$7020+3834i$	$-83277-56952i$	$i(1+i)^{36}3^{7}$	$3^5 73^3/2^3$	I*22,IV	$\bullet\!-\!\bullet\,{}^{3}$
142*(1+i)3	0	$621-135i$	$-4590+1512i$	$122472-64152i$	$2648943-4037688i$	$-i(1+i)^{48}3^{13}$	$3^5 17^3/2^9$	I*34,II*	

$141^*(1+i)3$	O	$-108+27i$	$-216-594i$	$3645+33534i$	$1537947+511515i$	$-i(1+i)^{36}3^{13}$	$3^5 73^3/2^3$	I*22,II*	•—• }3
$142^*i(1+i)$	O	$-207+972i$	$-510+510i$	$-300645-133218i$	$21100239-29321676i$	$i(1+i)^{48}3^7$	$3^5 17^3/2^9$	I*34,IV	
$141^*i(1+i)3$	O	$-108-27i$	$-216-216i$	$3645-33534i$	$1461402-598995i$	$i(1+i)^{36}3^{13}$	$3^5 73^3/2^3$	I*22,II*	•—• }3
142^*1+i	O	$207+972i$	$-510+510i$	$-300645+133218i$	$-21100239-29321676i$	$-i(1+i)^{48}3^7$	$3^5 17^3/2^9$	I*34,IV	

$$N=(1+i)^{12}(3)^5=(15552)$$

65^*1	O	O	O	$9i$	$-6-6i$	$-i(1+i)^{12}3^5$	$2^6 3^4$	II ,II	•
65^*i	O	O	O	$-9i$	$-6+6i$	$i(1+i)^{12}3^5$	$2^6 3^4$	II ,II	•
65^*3	O	O	O	$81i$	$-162-162i$	$-i(1+i)^{12}3^{11}$	$2^6 3^4$	II ,IV*	•
65^*i3	O	O	O	$-81i$	$-162+162i$	$i(1+i)^{12}3^{11}$	$2^6 3^4$	II ,IV*	•
65^*1+i	O	O	O	-18	24	$-i(1+i)^{18}3^5$	$2^6 3^4$	I*2 ,II	•
$65^*i(1+i)$	O	O	O	18	$-24i$	$i(1+i)^{18}3^5$	$2^6 3^4$	I*2 ,II	•
$65^*(1+i)3$	O	O	O	-162	648	$-i(1+i)^{18}3^{11}$	$2^6 3^4$	I*2 ,IV*	•
$65^*i(1+i)3$	O	O	O	162	$-648i$	$i(1+i)^{18}3^{11}$	$2^6 3^4$	I*2 ,IV*	•
122^*1	O	O	O	$27i$	$-36-36i$	$-i(1+i)^{12}3^7$	$2^6 3^5$	II ,IV	•

122^*i	o	o	o	$-27i$	$-36+36i$	$i(1+i)^{12}{}_3{}^7$	$2^6{}_3{}^5$	II ,IV	●
122^*3	o	o	o	$243i$	$-972-972i$	$-i(1+i)^{12}{}_3{}^{13}$	$2^6{}_3{}^5$	II ,II*	●
122^*i3	o	o	o	$-243i$	$-972+972i$	$i(1+i)^{12}{}_3{}^{13}$	$2^6{}_3{}^5$	II ,II*	●
122^*1+i	o	o	o	-54	144	$-i(1+i)^{18}{}_3{}^7$	$2^6{}_3{}^5$	I*2 ,IV	●
$122^*i(1+i)$	o	o	o	54	$-144i$	$i(1+i)^{18}{}_3{}^7$	$2^6{}_3{}^5$	I*2 ,IV	●
$122^*(1+i)3$	o	o	o	-486	3888	$-i(1+i)^{18}{}_3{}^{13}$	$2^6{}_3{}^5$	I*2 ,II*	●
$122^*i(1+i)3$	o	o	o	486	$-3888i$	$i(1+i)^{18}{}_3{}^{13}$	$2^6{}_3{}^5$	I*2 ,II*	●
124^*1	o	o	o	$81i$	$-198-198i$	$-i(1+i)^{12}{}_3{}^7$	$2^6{}_3{}^8$	II ,IV	●
124^*i	o	o	o	$-81i$	$-198+198i$	$i(1+i)^{12}{}_3{}^7$	$2^6{}_3{}^8$	II ,IV	●
124^*3	o	o	o	$729i$	$-5346-5346i$	$-i(1+i)^{12}{}_3{}^{13}$	$2^6{}_3{}^8$	II ,II*	●
124^*i3	o	o	o	$-729i$	$-5346+5346i$	$i(1+i)^{12}{}_3{}^{13}$	$2^6{}_3{}^8$	II ,II*	●
124^*1+i	o	o	o	-162	792	$-i(1+i)^{18}{}_3{}^7$	$2^6{}_3{}^8$	I*2 ,IV	●
$124^*i(1+i)$	o	o	o	162	$-792i$	$i(1+i)^{18}{}_3{}^7$	$2^6{}_3{}^8$	I*2 ,IV	●
$124^*(1+i)3$	o	o	o	-1458	21384	$-i(1+i)^{18}{}_3{}^{13}$	$2^6{}_3{}^8$	I*2 ,II*	●

$124^{*}i(1+i)3$	o	o	o	1458	$-21384i$	$i(1+i)^{18}3^{13}$	$2^{6}3^{8}$	I*2 ,II*	●

$$N=(1+i)^{13}(3)^{5}=(15552+15552i)$$

$67^{*}1$	o	o	o	$9-9i$	$-18-6i$	$-i(1+i)^{14}3^{5}$	$(1+i)^{13}3^{4}$	III ,II	●
$67^{*}i$	o	o	o	$-9+9i$	$-6+18i$	$i(1+i)^{14}3^{5}$	$(1+i)^{13}3^{4}$	III ,II	●
$67^{*}3$	o	o	o	$81-81i$	$-486-162i$	$-i(1+i)^{14}3^{11}$	$(1+i)^{13}3^{4}$	III ,IV*	●
$67^{*}i3$	o	o	o	$-81+81i$	$-162+486i$	$i(1+i)^{14}3^{11}$	$(1+i)^{13}3^{4}$	III ,IV*	●
$67^{*}1+i$	o	o	o	$18+18i$	$48-24i$	$-i(1+i)^{20}3^{5}$	$(1+i)^{13}3^{4}$	III*,II	●
$67^{*}i(1+i)$	o	o	o	$-18-18i$	$-24-48i$	$i(1+i)^{20}3^{5}$	$(1+i)^{13}3^{4}$	III*,II	●
$67^{*}(1+i)3$	o	o	o	$162+162i$	$1296-648i$	$-i(1+i)^{20}3^{11}$	$(1+i)^{13}3^{4}$	III*,IV*	●
$67^{*}i(1+i)3$	o	o	o	$-162-162i$	$-648-1296i$	$i(1+i)^{20}3^{11}$	$(1+i)^{13}3^{4}$	III*,IV*	●
$68^{*}1$	o	o	o	$9+9i$	$-18+16i$	$-i(1+i)^{14}3^{5}$	$-i(1+i)^{13}3^{4}$	III ,II	●
$68^{*}i$	o	o	o	$-9-9i$	$6+18i$	$i(1+i)^{14}3^{5}$	$-i(1+i)^{13}3^{4}$	III ,II	●
$68^{*}3$	o	o	o	$81+81i$	$-486+162i$	$-i(1+i)^{14}3^{11}$	$-i(1+i)^{13}3^{4}$	III ,IV*	●

$68^{*}i3$	o	o	o	$-81-81i$	$162+486i$	$i(1+i)^{14}3_3^{11}$	$-i(1+i)^{13}3_3^{4}$	III ,IV*	●
$68^{*}1+i$	o	o	o	$-18+18i$	$24-48i$	$-i(1+i)^{20}3_3^{5}$	$-i(1+i)^{13}3_3^{4}$	III*,II	●
$68^{*}i(1+i)$	o	o	o	$18-18i$	$-48-24i$	$i(1+i)^{20}3_3^{5}$	$-i(1+i)^{13}3_3^{4}$	III*,II	●
$68^{*}(1+i)3$	o	o	o	$-162+162i$	$648-1296i$	$-i(1+i)^{20}3_3^{11}$	$-i(1+i)^{13}3_3^{4}$	III*,IV*	●
$68^{*}i(1+i)3$	o	o	o	$162-162i$	$-1296-648i$	$i(1+i)^{20}3_3^{11}$	$-i(1+i)^{13}3_3^{4}$	III*,IV*	●

Table 4

Number of elliptic curves over $\mathbb{Q}(i)$ with conductor $N = (1+i)^a (3)^b$

a \ b	0	1	2	3	4	5	all b
0	0	0	0	4	0	4	8
1	0	0	0	18	12	12	42
2	0	0	4	6	8	8	26
3	0	6	6	4	8	10	34
6	2	4	8	6	4	12	36
7	0	4	4	4	8	8	28
8	2	14	22	42	40	54	174
9	8	8	24	24	16	4	84
10	12	36	68	108	96	112	432
11	0	0	0	16	0	0	16
12	8	24	40	24	24	24	144
13	16	16	32	48	32	16	160
14	16	0	32	32	16	0	96
all a	64	112	240	336	264	264	1280

Table 5

Number of isogeny classes of elliptic curves over $\mathbb{Q}(i)$ with conductor $N = (1+i)^a (3)^b$.

a \ b	0	1	2	3	4	5	all b
0	0	0	0	1	0	2	3
1	0	0	0	6	4	6	16
2	0	0	1	3	4	4	12
3	0	1	1	4	8	10	24
6	1	1	4	6	4	12	28
7	0	2	2	4	8	8	24
8	1	4	8	24	28	42	107
9	2	4	10	24	16	4	60
10	4	12	26	72	72	88	274
11	0	0	0	16	0	0	16
12	6	8	22	24	16	24	100
13	8	8	16	48	32	16	128
14	12	0	28	32	16	0	88
all a	34	40	118	264	208	192	880

References

[Ak] S. Akhtar, "Elliptic curves of prime power conductor", Punjab Univ. J. Math. (Lahore) 9 (1976), pp. 1-12. (see MR 58 # 27744).

[Ag&Coa&Hu&Po] M.K. Agrawal, J.H. Coates, J.C. Hunt, A.J. van der Poorten, "Elliptic curves of conductor 11 ", Math. Comp. 39 (1980), no. 151, pp. 991-1002.

[Ba 1] A. Baker, "Contributions to the theory of Diophantine equations II. The Diophantine equation $y^2 = x^3 + k$ ", Phil. Trans. Royal Soc. London, Series 4, 263 (1968), pp. 193-208.

[Ba 2] A. Baker, "Effective methods in Diophantine problems", Proc.Sympos.Pure Math., Amer. Math. Soc., Providence, R.I.,Vol.20, 1971, pp. 195-205;Vol.24, 1974, pp. 1-7.

[Bi&Swi 1] B.J. Birch, H.P.F. Swinnerton-Dyer, "Notes on elliptic curves I, II ", J. reine u. angew. Math. 212 (1963), pp. 7-25; 218 (1965), pp. 79-108.

[Bi&Swi 2]   B.J. Birch, H.P.F. Swinnerton-Dyer, "Elliptic curves and mo-
dular functions", Modular Functions of One Variable IV.
Lecture Notes in Math., Vol. 476, pp. 2-32, Springer-Verlag,
Berlin and New York, 1975.

[Bö]   R. Bölling, "Elliptische Kurven mit Primzahlführer", Math.
Nachr. 80 (1977), pp. 253-278.

[Bo&Sha]   S.I. Borewicz, I.R. Shafarevich, "Zahlentheorie", Birkhäuser
Verlag, Basel und Stuttgart, 1966.

[Bru&Kra]   A.Brumer, K. Kramer, "The Rank of Elliptic Curves", Duke
Math. J. 44 (1977), pp. 715-743.

[Ca]   J.W.S. Cassels, "Diophantine Equations with Special Referen-
ce to Elliptic Curves", J. London Math. Soc. Vol. 41 (1966),
pp. 193-291.

[Coa]   J. Coates, "An effective p-adic analogue of a theorem of
Thue", I, Acta Arithm. 15 (1969), pp. 279-305. II, "The
greatest prime factor of a binary form", Acta Arithm. 16
(1970), pp. 399-412. III, "The diophantine equation
$y^2 = x^3 + k$", Acta Arithm. 16 (1970), pp. 425-435.

[Cog 1]   F.B. Coghlan, "Elliptic curves with conductor $N=2^m3^n$ " ,
               Ph.D. thesis, Manchester University (1967). Tables pub-
               lished in: Modular Functions of One Variable IV. Springer
               Lecture Notes in Math. 476 (1975) pp. 123-134. Springer
               Verlag.

[Cog 2]   F.B. Coghlan, "The Diophantine Equation $x^3-y^2=k$ " , Com-
               puters in Number Theory. (Oxford 1969), ed.: Atkin and
               Birch. Ac. Press. 1971.

[Dem]   V.A. Demyanenko, "Orders of the torsion points of curves
               of genus 1 " , J. Sov. Math. 18, 843-861 (1982).

[Deu]   M. Deuring, "Die Typen der Multiplikatorenringe ellip-
               tischer Funktionenkörper", Abh. Math. Sem. Hamburg, 14
               (1941), pp. 197-272.

[El&Gru&Me]   J. Elstrodt, F. Grunewald, J. Mennicke, "On the group
               $PSL_2(\mathbf{Z}[i])$", Proc. Conf. Journées Arithmetiques, Exeter,
               1980, pp. 255-283.

[Ge]   St. Gelbart,"Elliptic curves and automorphic representa-
               tions", Advances in Math. 21 (1976, pp. 235-292.

[Gru&Me]   F.J. Grunewald, J. Mennicke, "$SL_2(O)$ and elliptic curves",
               Preprint, Bielefeld 1978.

[Ha 1]   T. Hadano, "On the conductor of an elliptic curve with a
           rational point of order 2", Nagoya Math. J. 53 (1974),
           pp. 199-210.

[Ha 2]   T. Hadano, "Elliptic curves with a rational point of finite
           order", Manuscr. Math. 39 (1982), pp. 49-79.

[Is]   H. Ishii, "The nonexistence of elliptic curves with every-
           where good reduction over certain imaginary quadratic
           fields",J. Math. Soc. Japan 31 (1979), no.2, pp. 273-279.

[Kam]   S. Kamienny, "Points of order p on elliptic curves over
           Q $(\sqrt{p})$". Math. Ann. 261 (1982), pp. 413-424.

[Ke 1]   M.A. Kenku, "The modular curve $X_0(39)$ and rational isogeny",
           Math. Proc. (Cambr. Philos. Soc. 85(1979), no. 1,pp.21-23.

[Ke 2]   M. A. Kenku, "The modular curve $X_0(169)$ and rational iso-
           geny", J. London Math. Soc. (2), 22 (1980), no. 2,
           pp. 239-244.

[Ke 3]   M. A. Kenku, "The modular curves $X_0(65)$ and $X_0(91)$ and
           rational isogeny", Math. Proc. Cambridge Philos. Soc. 87
           (1980), no. 1, pp. 15-20.

[Ke 4]   Corrigendum:"The modular curve $X_o(169)$ and rational iso-
             geny" [J. London Math. Soc. (2) 22(1980), no.2, pp.239-244],
             J. London Math. Soc. (2) (23) (1981), no. 3, p. 428.

[Lan 1]   S. Lang, "Elliptic Functions", Addison Wesley, London 1973.

[Lan 2]   S. Lang "Elliptic Curves. Diophantine Analysis", Grundleh-
             ren der mathematischen Wissenschaften 231, Springer Verlag,
             Berlin, Heidelberg and New York, 1978.

[Las 1]   M. Laska, "An algorithm for finding a minimal Weierstrass
             equation for an elliptic curve", Math. Comp., Vol. 38,
             no. 157, 1982, pp. 257-260.

[Las 2]   M. Laska, "Die diophantische Gleichung $x^3-y^2=\varepsilon(1+i)^m 3^n$
             über $\mathbf{Z}[i]$ und die Klassifikation gewisser elliptischer
             Kurven über $\mathbb{Q}(i)$", Dissertation, Bielefeld, 1980.

[Las 3]   M. Laska, "Solving the equation $x^3-y^2=r$ in number fields",
             J. reine u. ang. Math. 333 (1982), pp. 73-85.

[Las 4]   M. Laska, "Elliptic curves over $\mathbb{Q}(i)$ with conductor
             $N=(1+i)^a (3)^b$ ", Preprint Sonderforschungsbereich 40, Theo-
             retische Mathematik, Bonn 1982.

[Las 5]   M. Laska, "Determining binary cubic forms over number rings
          with given discriminant", in preparation.

[Li]   G. Ligozat, "Courbes elliptiques ayant bonne réduction en
          dehors 3 ", Nagoya Math. J. 76 (1979), pp. 105-152.

[Mah]   K. Mahler, "Ueber die rationalen Punkte auf Kurven von Ge-
          schlecht 1", J. reine u. ang. Math. 170 (1933), pp.168-178.

[Mar]   D.A. Marcus, "Number fields", Springer Verlag, Berlin,
          Heidelberg and New York, 1977.

[Maz 1]   B. Mazur, "Modular curves and the Eisenstein ideal", Publ.
          Math. I.H.E.S., 47 (1978), pp. 33-186.

[Maz 2]   B. Mazur, "Rational isogenies of prime degree", Inv. Math.
          44 (1978), pp. 129-162.

[Mi]   I. Miyawaki, "Elliptic curves of prime power conductor with
          Q-rational points of finite order", Osaka J. Math. 10 (1973),
          pp. 309-323.

[Mo]   L.J. Mordell, "Diophantine Equations", Academic Press, Lon-
          don and New York, 1964.

[Neu 1]   O. Neumann, "Zur Reduktion der elliptischen Kurven",
            Math. Nachr. 46 (1970), pp. 258-310.

[Neu 2]   O. Neumann, "Die elliptischen Kurven mit den Führern
            $3 \cdot 2^m$ und $9 \cdot 2^m$", Math. Nachr. 48 (1973), pp. 387-389.

[Neu 3]   O. Neumann, "Elliptische Kurven mit vorgeschriebenem Reduk-
            tionsverhalten I, II", Math. Nachr., Bd. 49 (1971),
            pp. 107-123, Bd. 56 (1973), pp. 269-280.

[Ogg 1]   A.P. Ogg, "Elliptic curves and wild ramification", Amer. J.
            Math. 89 (1967), pp. 1-21.

[Ogg 2]   A.P. Ogg, "Abelian curves of 2-power conductor", Proc. Cambr.
            Phil. Soc. 62 (1966), pp. 143-148.

[Ogg 3]   A.P. Ogg, "Abelian Curves of Small conductor", J. reine u.
            ang. Math. 226 (1967), pp. 204-215.

[Ogg 4]   A.P. Ogg, "Rational points on certain elliptic modular
            curves", Proc.Symp. Pure Math. AMS Providence, 24 (1973),
            pp. 221-231.

[Ro]   A. Robert, "Elliptic Curves", Lecture Notes in Math.,
            Vol. 326, Springer Verlag, Berlin and New York, 1973.

[Ser]   J.P. Serre, "Abelian ℓ-Adic Representations and Elliptic
            Curves", W.A. Benjamin, New York, Amsterdam 1968.

[Ser&Ta]   J.P. Serre, J. Tate, "Good reduction of abelian varieties",
            Ann. of Math. 88 (1968), pp. 492-517.

[Set 1]   B. Setzer, "Elliptic curves of prime conductor", J. London
            Math. Soc. (2), 10 (1975), pp. 367-378.

[Set 2]   B. Setzer, "Elliptic curves over complex quadratic fields",
            Pacific J. Math., Vol. 74 (1978), pp. 235-250.

[Set 3]   B. Setzer, "Elliptic curves with good reduction everywhere
            over quadratic fields and having rational j-invariant",
            Ill.J. Math. 25 (1981), pp. 233-245.

[Si]   C.L. Siegel, "Über einige Anwendungen diophantischer Approxi-
            mationen", Abhandlungen der Preußischen Akademie der Wissen-
            schaften. Physikalisch-mathematische Klasse 1929, Nr. 1.

[Stro 1]   R.J. Stroeker, "Aspects of elliptic curves; an introduction",
            Nieuw Archief voor Wiskunde (3), XXVI (1978), pp.371-412.

[Stro 2]   R.J. Stroeker, "Elliptic curves defined over imaginary
            quadratic number fields", Ph. D. thesis, Amsterdam 1975.

[Ta 1]   J.T. Tate, "The arithmetic of elliptic curves", Inv. Math. v. 23, 1974, pp. 179-206.

[Ta 2]   J.T. Tate, "Algorithm for determining the type of singular fiber in an elliptic pencil", Modular Functions of One Variable. IV. Lecture Notes in Math., Vol. 476, pp. 33-52, Springer Verlag, Berlin and New York, 1975.

[Ve 1]   J. Velu, "Isogénies entre courbes elliptiques", C.R. Acad. Sc. Paris 273(1973), pp. 238-241.

[Ve 2]   J. Velu, "Courbes elliptiques sur $\mathbb{Q}$ ayant bonne reduction en dehors de {11}" , C.R. Acad. Sci. Paris 273 (1971), pp. 73-75.

Index of Special Symbols

b_2 , b_4 , b_6 , b_8	see 1.2
c_4 , c_6	see 1.3
$\Delta(\Gamma)$	discriminant of Γ , see 1.3
j , $j(E)$	j-invariant, see 1.3
$\Gamma^{(u)}$	see 1.5
$\mathfrak{p}^{d_\mathfrak{p}}$, $\mathrm{Disc}(E)$	discriminant at $\mathfrak{p}$, discriminant of E
$E_\mathfrak{p}$	reduction of E at $\mathfrak{p}$
$E_\mathfrak{p}(k_\mathfrak{p})$, $E_\mathfrak{p}(k_\mathfrak{p})^{ns}$	set of $k_\mathfrak{p}$-rational, non-singular $k_\mathfrak{p}$-rational points on $E_\mathfrak{p}$
$a_\mathfrak{p}$	$\|\mathfrak{p}\| + 1 - \|E_\mathfrak{p}(k_\mathfrak{p})\|$
$L(E,s)$	L-series of E
X	Néron model
$X_\mathfrak{p}$	Néron-Kodaira reduction at $\mathfrak{p}$
$n_\mathfrak{p}$	number of irreducible components of $X_\mathfrak{p}$
$\mathfrak{p}^{f_\mathfrak{p}}$, $\mathrm{Cond}(E)$	conductor at $\mathfrak{p}$, conductor of E
$\mathcal{E}(S)$	set of elliptic curves over K with good reduction outside S
B $(=B_S)$	fixed set of pairwise non-associated basic solutions of equation (*) in chapter 2
U $(=U_S)$	$0_K^{*}\langle S\rangle$
$\beta*\overline{\lambda}$, $\beta*\lambda$	see Theorem 2.4 and the remark preceding Theorem 2.7
$\Gamma_{v,w}$	see the discussion preceding Lemma 2.6
$\mathfrak{R}$	$\{\zeta^\mu{}_n{}^\nu \pi_1^{\alpha_1}\cdots\pi_n^{\alpha_n} \mid \mu,\nu,\alpha_1,\ldots,\alpha_n = 0 \text{ or } 1\}$

Γ_r	equation $x^3 - y^2 = r$
$\Gamma_r(R)$	set of solutions of Γ_r in $R \times R$
Ξ	see Lemma 3.2
$\amalg_{r,L}$, $\amalg_r$	see 3.6 and the subsequent discussion
$\mathfrak{S}_{r,L}$, $_r$	see 3.6 and the subsequent discussion
$V(F)$	see 3.6
$f_{\sigma,\tau}$, $g_{\sigma,\tau}$	see 3.7
$h_{\sigma,\tau}$	see 3.11
K_r	see the discussion after 3.12
$V(f_{\sigma,\tau})$	see the discussion preceding assumption 3.13
E_r	elliptic curve defined by Γ_r
σ_r	"trace map" from $E_r(K)$ to $E_r(\mathbb{Q})$
$\sigma_{r,R}$	map from $\Gamma_r(R)$ to $\Gamma_r(\mathbb{Q}) \cup \{\underline{0}\}$, induced from σ_r
$E(L)$	group of L-rational points on a fixed Weierstrass equation for E
$M^{\#}$, M_2 , M_x	see the beginning of chapter 4
$E(K)_{tors}$	torsion subgroup of the Mordell-Weil group
$D_n(E)$	set of n-division points in $E(\mathbb{C})$
f_n	see 4.2
$T_n(E)$	n-division polynomial of E
$K(D_n(E),E)$	n-division field of E
$G_n(E)$	Galois group of $K(D_n(E))$ over K
R_n	faithful representation of $G_n(E)$ in $GL_2(\mathbb{Z}/n\mathbb{Z})$
g_n , h_n	see 4.3
H_U	subgroup polynomial corresponding to the subgroup U
$K(U)$	extension field of K generated by the x- and y-coordinates of the elements in $U^{\#}$

$\underline{U}_n(E)$ set of subgroup polynomials of order n of E with coefficients in K .

$\lambda_{L'}$ homomorphism of the groups of L'-rational points, induced from λ

$\underline{I}_n(E)$ set of elliptic curves over K , n-isogenous to E

E/U see remarks after Proposition 4.2o

Ψ_n see Corollary 4.23

$\underline{I}_n^{cyc}(E)$ see 4.17

Index